Modelling Large Systems

Modelling Large Systems

Peter C. Roberts

ORASA text No. 4

Taylor & Francis · London · 1978

First published 1978 by Taylor & Francis Ltd,
10–14 Macklin Street, London WC2B 5NF

ISBN 0 85066 170 6

Printed and bound in the United Kingdom by
Taylor & Francis (Printers) Ltd, Rankine Road,
Basingstoke, Hampshire RG24 0PR.

Distributed in the United States and its Territories
by Halsted Press (a division of John Wiley & Sons Inc.)
605 Third Avenue, New York, N.Y. 10016.

Contents

Introduction

There are few phrases that give rise to such emotion in the world of the traditional academic or scientific adviser than does 'world modelling'. Born of contentious argument, fostered by naive and simplistic propaganda, the subject lives on in a world of exaggeration and partisanship. It is a debate that has forced major issues into public awareness when more traditional analysts would have suppressed them entirely, but on the whole neither side has emerged from the debate with much credit.

One of the revealing features of the debate has been the way in which so many of the contenders have revealed their ignorance of the fundamental principles of systems modelling—an ignorance which seems shocking until one tries to find a reasoned statement of those principles. There is a serious gap in the literature at this point and this book is an attempt to fill part of that gap. It is written by the Head of the Systems Analysis Research Unit of the Department of the Environment in the UK, which is one of the world's leading research teams in this area. The Unit is also unique in that it has approached its work without a political axe to grind, and started with a careful review of what had already been done before developing its own models.

The text is introductory—and should perhaps be made compulsory reading for anybody wishing to express an opinion on a world model, let alone build one. Its foundation is that of traditional science, and shows how scientific ideas develop naturally into work of this kind. But it also shows the constraints and conditions necessary for the development of a satisfactory model. It started out to be an explanatory text, but has become much more.

ROLFE TOMLINSON
General Editor

Preface

This is a book about world models though less than half of it deals directly with past or potential world models. The reason for this lies in the fact that world modelling brings into focus all of those deep seated problems of models concerned with credibility, verification, forecasting and 'who changes what' in real situations. Discussions about world modelling tend to generate much heat. I have attempted to shed some light in these pages, by examining the nuts and bolts of the modelling process.

I make no apology for extending the field of inquiry to the frontier with philosophy. All models, not only world models, force one to probe the philosophical underpinning. The natural sciences are replete with models of all sorts and the illustrative material for the early chapters draws heavily on these. I have argued that operational researchers should not forget their origins in the more traditional sciences.

My thanks are due to past and present members of the Systems Analysis Research Unit, who contributed to that peculiarly astringent style of critical discussion which forces the rejection of ill-founded assertions.

1977 PETER ROBERTS

Acknowledgements

My thanks for permission to reproduce illustrations are due to:

Blackwell Scientific Publications for Figure 2.1 from *Theoretical Ecology*, May 1976.

Wright Allen Press Inc for Figures 2.2, 2.3, 2.4, 2.5 and 2.6 from Meadows, D. L., 1970, Dynamics of Commodity Production and Cycles; for Figure 5.4 from Forrester, J. A., 1971, *World Dynamics*.

John C. Fisher for Figure 4.1 from Fisher, J. C. & Pry, R. H., June 1970, *A Simple Substitution Model of Technological Change*, Report GE 70-C-25.

F. Meyer for Figures 5.1, 5.2 and 5.3 from Meyer, F. and Vallee, J., 1975, *The Dynamics of Long Term Growth*, Technological Forecasting and Social Change 7 (285-300).

1. Science and systems

Throughout recorded history writers have sought the marks that distinguish human kind from beasts of the field. For the religious it is a simple distinction between those with and those without immortal souls. In times noted by a stronger taste for the empirical the distinction was 'man the toolmaker'. More appealing is 'man the artist', and for a touch of whimsy, *homo ludens*—'man the games player'. No doubt there are others and they collectively illustrate a higher level, an all embracing distinction, *homo generalis*—'man the generalizer'; the categorizer, botanist, propounder of natural law, seeker for similarity amidst the differences, and unity within diversity. In one way or another we all enjoy the process of reductionism, whereby the complexity of nature is explained by one general principle. A frisson of pleasure and awe runs through us at the idea that all the wealth of diversity in the physical world is created out of a handful of elements, or that the primates have topologically identical bone structures, or that all plane right-angled triangles are governed by the Pythagorean rule (including all the plane right-angled triangles that could ever be drawn).

As usual, snobbery and elitism abound. Rutherford, in his usual pithy style, wrote on the lower orders of the scientific world: 'There is physics, and there is chemistry (which is an inferior form of physics)—and there is stamp collecting'. You may not like what he says, but you know what he means. If Rutherford could claim a place in the scientific aristocracy for physics because of the generality of its statements, we should award honourable mention to the system scientists. Their generalizations are of function, of

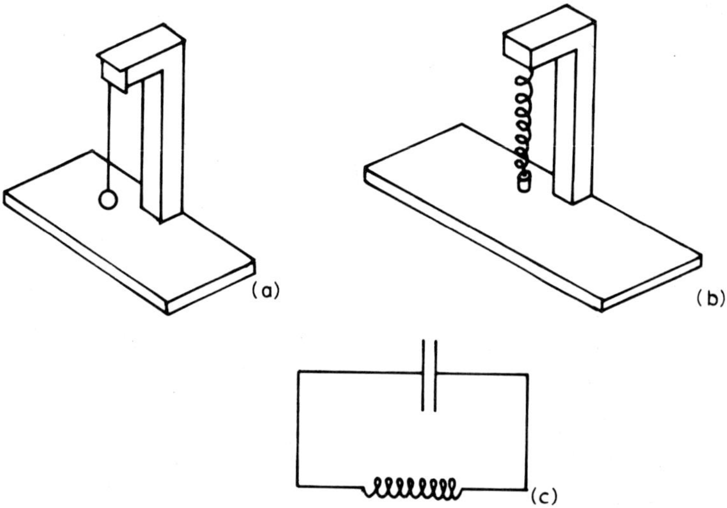

FIGURE 1.1 Three forms of oscillatory system: pendulum, vertical spring with suspended mass, series circuit of inductor and capacitor.

modes of behaviour, of dynamic properties, but their interest is in the structure, not the stuff. Consider the systems of Figure 1.1:

(*a*) Pendulum.
(*b*) Vertical spring with suspended mass.
(*c*) Series circuit of inductor and capacitor.

The materials are very different, the parameters that determine behaviour are specific to each example, but at a more abstract level these systems belong to the same class. Ignore damping and a single differential equation can be used to define the motion (whether of pendulum bob, suspended mass or current in the circuit).

Thus,

$$\frac{d^2x}{dt^2} = -kx \qquad (1.1)$$

where x is a measure of the displacement from equilibrium such that:

For the pendulum, x = angle of separation from the vertical (for small angles, say $< 5°$).
For the spring, x = vertical distance of the mass from the rest position.
For the circuit, x = potential difference between the plates of the capacitor; t = time; k = a positive constant unique to each system, but always with the dimensions of the inverse square of time.

It can be shown that the differential equation integrates to yield a relation of the form:

$$x = a \sin \omega (t + \varphi) \tag{1.2}$$

where $\omega = 1/\sqrt{k}$ and a, φ are integration constants. This is the equation of an oscillator—a sine wave: repetitive cycles in which the sequence of x values recurs again and again every $2\pi/\omega$ units of time.

Though simple in character, this equation and its solution exemplifies that feature of systems which provokes interest. The component parts—the bob, string, spring, mass, capacitor, inductor—are none of them intrinsically oscillatory; but arranged in suitable relationships, the combinations of components can possess oscillatory properties. For larger systems, those with many component parts, this idea is brought out by contrasting the micro behaviour (that of the components) with the macro behaviour (that of the total system), Thus, we may contrast the chaotic dance of the gas molecules inside a balloon (the micro aspect) with the apparently steady and uniform outward force per unit area keeping the balloon inflated (the macro aspect). At the biological level, the outward manifestations of muscular contraction derive from changes to the concentrations of ions present in the cells. Hence the concept of reducing the elaborate behaviour of biological entities to the chemistry of their component molecules, and thence to the physics rules governing which molecules are formed from given elements.

A more evocative term has been coined for the systems that we call life forms: 'emergent properties'. Those qualities we associate with life, like self-replication, are regarded as emerging inevitably from large but dull amino acid molecules. It has to be said that the virus or even the amoeba emerging in this way does not challenge credibility, but that it takes an ardent reductionist to believe, say, in Bach's violin concerto in A minor 'emerging' from a mass of protoplasm.

The newcomer could be forgiven for thinking that systems science, systems engineering, systems analysis and cybernetics were brand-new shiny products of the second half of the twentieth century, since it is only in that period that the terms have come into common use. The truth is, however, that important ideas in systems thinking have a long pedigree, and some of the most striking and successful results of the systems approach belong to classical science. Indeed, so heavily dependent are many modern applications on classic models that it is worthwhile tracing their precedents, if only to appreciate that a few profound insights by the early giants can spawn generations of model builders in emulation.

Of all the classical examples which could be used, the story of the development of models for understanding the solar system is the most striking, in the sense it conveys of successive penetration to ever simpler components or principles from which the variety of outward behaviour emerges: a reduction further and further back to a most satisfying and elegant synthesis. In its last, or perhaps one should say 'current' form, the application of the system rules is called 'celestial mechanics'. This has a prosaic flavour compared with the 'music of the spheres', but it would not be inapposite to describe the end result as a 'divine simplicity'.

Celestial mechanics

The evolution in understanding of planetary motion has been of interest to historians of politics, religion and scientific thought.[1] Those planets visible to the naked eye (Mercury, Venus, Mars, Jupiter and Saturn) have apparent motions which are perplexing. For most of the time they move against the background of fixed stars in stately regular courses, but from time to time each one slows and then makes a short, retrograde step followed by a further reversal and return to forward motion. The first quantified explanation for this effect of which we are aware is due to Ptolemy in about A.D. 100. The ideas of reductionism and simple underlying principles are there, based on an idea from Aristotle who had declared that perfect motion is circular motion. So the Ptolemaic system is an ingenious mix of epicycles and deferents, circles upon circles which, given a geocentric system yields a rough approximation to the planets' motions including the retrograde phases. The heliocentric system (offered speculatively by Anaxogoras) was revived by Copernicus with the advantage of explaining retrograde motion much more elegantly, but Kepler took the next crucial step in abandoning the circles for ellipses. The Kepler laws of planetary motion give very good agreement with observation. Up to this point we see descriptions of the system with higher accuracy and using fewer parameters, but low in explanatory power (a mere five planets). Newton postulated a breath-taking generalization that not only did the elliptical orbits follow from a rule of attractive force between sun and planet varying inversely with the square of the intervening distance, but that there was a root cause of this, namely that every particle attracts every other particle according to the product of the particle masses, and inversely as the square of the particle separation. This exquisitely simple proposition has tremendous explanatory power: the motions of planets, comets, moons; tidal variation; oblateness of the earth spheroid, precession of the equinoxes and variation of pendulum swing with altitude–latitude. The remarkable success which attended this essay, to reduce a great diversity of celestial and terrestial phenomena to a universal underlying law, has encouraged three centuries of scientists and a quarter of a century of operational researchers and systems analysts to do likewise.

Urban mechanics 1

The number of objects larger than, say, a mile in diameter, that occupy the space that we call the solar system is large. The main features of the motions of these objects are puzzling, and the finer features baffling, without the guidance of the gravitational principle. But if such a unification as the Newtonian law can be demonstrated for one large system, there is reason to hope that similar feats could be achieved for others. One of the best known recent system applications following the same basic concept has been to the traffic flows in urban areas.

The traffic flows in a modern city are on a scale which certainly qualify for description as a 'large system'. In orders of magnitude there are 10^6 people, 10^4 links and nodes and 10^9 trips made in a large city each year. The capital investment in roads, vehicles, track, tunnels, bridges, etc. is a significant proportion of total public investment; the running costs in fuel, lighting, maintenance, policing and hospital care are at a similar level, and the time outlay of travellers is perhaps a third of their disposable hours (time not used in working, sleeping and eating). A set of transportation modellers has, therefore, come into being with the mission of understanding and explaining the properties of the system. This is not just an academic exercise: at any rate, the transportation modellers in government service have the goal of estimating the cost effectiveness of changes to the transport system.

People are not particles, and the 'forces' that drive them hither and thither are not gravitational in nature; nevertheless, gravity provides a useful analogy on which to build a transportation model. Suppose the ground area of the city is divided into small 'zones'. It is not unreasonable to assume that the number of trips made between zone i and zone j would be inversely related to the difficulty of getting from i to j. Although distance_{ij} may be a rough measure of difficulty, a preferred variable is cost_{ij}, and because people value time as well as money, cost of getting from zone i to zone j should take account of both the monetary expenditure and the perceived value of travel time (and conceivably, the cost of dis-discomfort). Nothing can yet be said about the 'deterrence function', the manner in which the number of trips_{ij} depends upon the generalized cost_{ij}—certainly we have no right to assume a square law dependence. Nevertheless, we can see that this law has the right pattern of behaviour, providing steadily increasing deterrence as cost increases.

It is easy to see the other constituents of the transportation model as being analogous to mass. The floor area of the buildings in the zone has been proffered as one possible analogue to mass. A more ingenious way of securing the analogue is to argue that the number of trips from source i to destination j is proportional to the total number of trips with source i to all other zones. Similarly, it can be hypothesized that the number of trips with source i and destination j is proportional to the total number of trips with destination j starting from all other zones, including i. These innocent, and even apparently tautologous assumptions, have far reaching implications. Expressed in mathematical form we have:

Let

T_{ij} = number of trips source i, destination j,
c_{ij} = generalized cost of trip between i and j,
$\sum_j T_{ij} = O_i$ total number of trips source i to all destinations,
$\sum_i T_{ij} = D_j$ total number of trips destination j starting from all origins,

and let $f(c_{ij})$ be a deterrence function.
Then

$$T_{ij} = A_i B_j \frac{O_i D_j}{f(c_{ij})} \tag{1.3}$$

Compare this and the equation for the gravitational attraction F between two point masses m_i and m_j separated by a distance d:

$$F_{ij} = \frac{G\, m_i m_j}{d^2{}_{ij}} \tag{1.4}$$

In this expression, one constant only is needed for all i, j—the gravitational constant G. In the transportation expression, on the other hand, many constants A_i and B_j are required specific to the contexts of i and j respectively. The reason for this modification, rather than using a single universal constant, can be understood by imagining an increase of O_i to $O_i \times 10$ and of D_j to $D_j \times 10$. Clearly the resulting T_{ij} would not be as large as $\times$ 100 compared with the original T_{ij}. Such a manner of T_{ij} increase would mean that total trips among all zones was dependent on the boundaries chosen for the zones, and this choice is arbitrary.

Before developing the transportation model further, it will be helpful to consider another piece of invention from classical science.

Gas Mechanics

A gas consists of a large number of molecules moving about and rebounding from each other and from the walls of the containing vessel. If it is assumed that these motions are random, that there is no preferred direction of motion for any particular molecule at any specified time this provides a sufficient specification to define a model from which general deductions can be made using probability methods. In Appendix 1 the expressions for the distribution of molecular density and of molecular energy are derived. It appears that the most likely distribution of molecules is that in which equal volumes within the vessel contain equal numbers of molecules, the density tends to be uniform throughout the vessel. This is an unsurprising result but if the same technique is applied to the *energies* of the molecules, an unexpected result appears. In fact the most likely distribution is one in which the numbers of molecules in equi-width energy bands, of successively higher mean energy, declines geometrically. A plot of numbers of molecules against energy of those molecules is negative exponential in form. This result is far from obvious and quite elaborate experimental means has been needed to verify that, indeed, the energy distribution of real gases conforms to this rule.

Urban mechanics 2

This technique of modelling a system by deducing the most probable amongst all its possible states has prompted A. G. Wilson[2] to devise a similar approach for transportation. In Appendix 2 it is shown that a treatment of trip making, strictly parallel with that used to model gas molecules, can be used to identify the form of the deterrence function. The two constraints applying in the gas system are:

1 Fixed number of molecules present;
2 Fixed total energy of the molecular ensemble.

The constraints applying to the urban system are, in the nomenclature of Urban Mechanics 1:

1 $\sum_j T_{ij} = O_i$ for all origins;
2 $\sum_i T_{ij} = D_j$ for all destinations;
3 $\sum_i \sum_j T_{ij}c_{ij} = C$ the total money spent on all tripmaking.

The deterrence term emerges as an exponential function of the generalized cost.

$$f(c_{ij}) = A\, e^{\lambda c_{ij}} \qquad (1.5)$$

where A and λ are constants particular to the city and its inhabitants. This is in clear contrast with the Newtonian gravity denominator, which is the square of separating distance. That such a specific deterrence function should emerge from such simple ideas is remarkable because the function is, from a different viewpoint, descriptive of aggregate human behaviour. However, there is an alternative way of thinking about trip distributions which also leads to a similar deterrence function. Assume that each trip is made to achieve some objective. Given a uniform density of locations where the objective can be achieved, then for each equal increment of distance travelled Δx, let there be equal probability ρ of terminating the trip successfully, so that for a distance L there is a probability $(1-\rho)^{L/\Delta x}$ of failing to be successful at all intervening points. Thus the proportion made of long trips L_1 will be smaller than that for short trips L_2 and the ratio between them will be

$$(1-\rho)^{L_1/\Delta x} \div (1-\rho)^{L_2/\Delta x} = (1-\rho)^{(L_1-L_2)/\Delta x} \qquad (1.6)$$

Assuming that cost is proportional to distance travelled, then the form of the relation between number of trips and cost will be:

$$T_{ij} = R(1-\rho)^{-\lambda c_{ij}}$$

R and λ constants

$$= Re^{-a\lambda c_{ij}} \tag{1.7}$$

where $a = \log(1-\rho)$.

This reasoning chain underlies the intervening-opportunities model. It contains heroic assumptions but the agreement of form for the deterrence function is intriguing.

Ideally one would wish to test these models by determining, say, the actual costs involved. Unfortunately in practice it is difficult to establish a good set of interzonal costs (particularly where different methods of travel exist), and to separate one type of trip (like journey-to-work). Nevertheless, there is accumulating evidence that the exponential deterrence function provides a good fit for travel patterns in an urban environment.

Here then, we start with the strong indication that the same kind of generalizing laws on which progress in the physical sciences has been based may, in some cases at least, also be applied to problems of social behaviour and national life.

2. Natural ripples

Modelling populations of plants and animals presents a greater challenge than finding good representations of molecular behaviour. Naturally occurring systems of living organisms are rich in complexity and the search for regularities is a tantalizing one. Here and there, the synthesis from simple elements to yield emergent patterns corresponding with the observed dynamics is possible. Ecology is a relatively new science and it is premature to ask for a set of general laws. However, there are already some interesting examples of possible modelling approaches.

Single populations, although these are rare in natural systems, enable some simple ideas to be developed. If the *per capita* growth rate r is independent of population density, a simple differential equation can be written:

$$\frac{dN}{dt} = rN \qquad (2.1)$$

where N is the population.
This has the solution:

$$N_t = N_0 e^{rt} \qquad (2.2)$$

For $r > 0$ this yields unbounded growth which is not to be found in nature. A simple model which captures the essential features of a finite environment is the logistic equation:

$$\frac{dN}{dt} = rN(1 - N/K) \qquad (2.3)$$

Here the effective *per capita* growth rate is dependent on the population and has the form $r(1-N/K)$ which is positive if $N < K$ and negative if $N > K$; this leads to a stable equilibrium population. K may be thought of as the carrying capacity of the environment determined by food, space, predators, etc. and r as the 'intrinsic' growth rate from environmental constraints. In Equation 2.3 the density dependent regulatory mechanism represented by the factor $(1-N/K)$ operates instantaneously. In most real life situations, these regulatory effects are likely to operate with some built-in time lag, whose characteristic magnitude may be denoted by T. Such time lags may, for example, derive from vegetation recovery times or other environmental effects; or from the time of approximately one generation which elapses before the depression in birth rates at high densities shows up as a decrease in the adult population. A crude way of incorporating such time delays is to rewrite Equation 2.3 as:

$$\frac{dN}{dt} = rN[1-N(t-T)/K] \tag{2.4}$$

The qualitative nature of the solutions of this equation follow from precepts familiar to engineers. If the time delay T in the feedback mechanism is long compared to the natural response time of the system $1/r$, there will be a tendency to overshoot and to overcompensate. For modest values of the time delay this overcompensation produces an oscillatory, rather than a monotonic, return to the equilibrium point. As the time delay becomes longer (as rT exceeds some number of order unity) there is a so-called Hopf bifurcation, and the stable point gives way to stable limit cycles. These stable limit cycles are an explicitly non-linear phenomenon, in which the population density oscillates up and down in a cycle whose amplitude and period is determined uniquely by the parameters of the equation. Just as in the case of a stable equilibrium point, if the system is perturbed it will tend to return to this stable cyclic trajectory.

Growth, in accord with the logistic Equation 2.3, has been shown in experiments using laboratory populations in constant and limited environments. For the protozoan *Paramecium*, on yeast cells, on bacteria such as *E. coli*, on *Drosophila* and on various flour beetles, logistic growth has been exhibited.

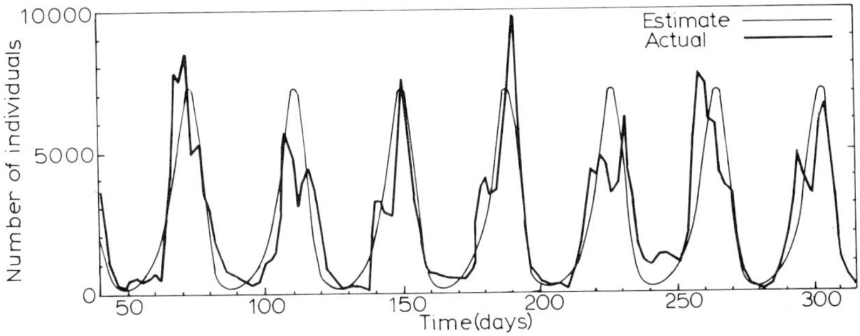

FIGURE 2.1 Comparison of Nicholson's study of the life-cycle of the sheep-blowfly (*Lucilia cuprina*) with the observed results.

Other similar single species experiments have shown sustained patterns of large-amplitude oscillations. The most notable of these is Nicholson's[3] study of the sheep-blowfly *Lucilia cuprina*. An accurate model for this population would need to include both the details of the reproductive physiology of blowflies, and the fact that there are age classes. However, the essential dynamics of the system may be captured by the naive Equation 2.4 which includes three major features: an intrinsic rate of increase r; a resource limitation K, set by the amount of a constant supply of ground liver; and a time delay T, in the action of this limitation roughly equal to the time for a larva to mature into an adult. Figure 2.1 shows the fit obtained using a single parameter rT to match the observed ratio between maximum and minimum populations. From this particular choice of rT, one can infer an egg-to-adult development time of 9 days, in fair agreement with the actual figure of 11 days. Recalling the gross simplifications used in setting up this model, the concordance of observed wave form with that dictated by the theory is striking.

Commodity cycles

It seems a far cry from blowflies feeding on liver to production-consumption systems in human societies, though for the systems analyst the identification of structural similarities can take place across a wide range of contexts. The feedback mechanisms present in the response of animal populations are analogous to the feedbacks observable in economic systems. One study due to D. L. Meadows[4] on commodity cycles shows the parallel strikingly. The feedback loop structure for a commodity is drawn in Figure 2.2.

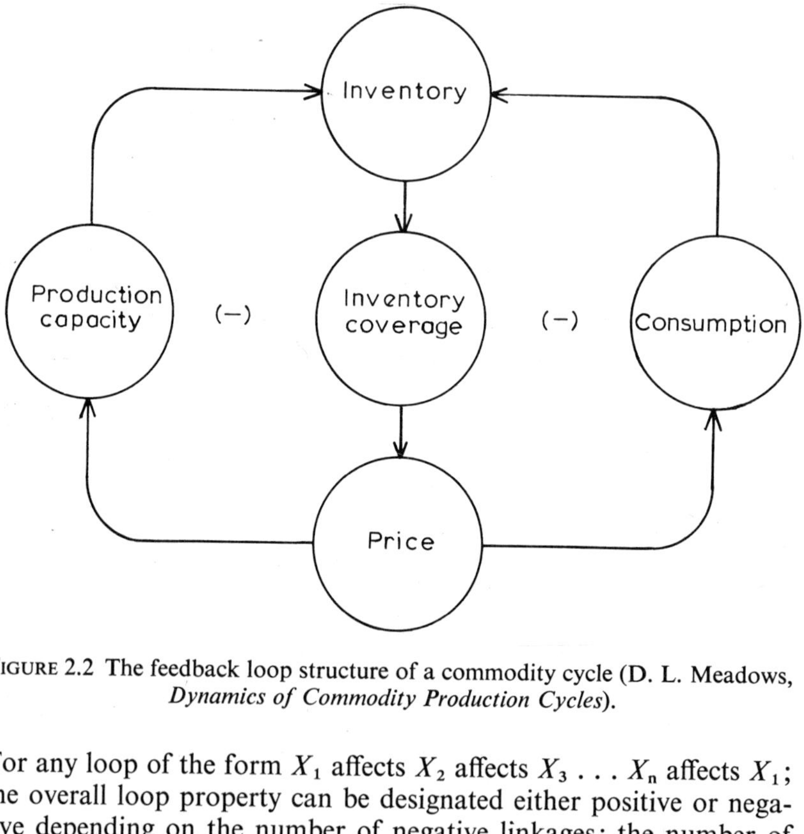

FIGURE 2.2 The feedback loop structure of a commodity cycle (D. L. Meadows, *Dynamics of Commodity Production Cycles*).

For any loop of the form X_1 affects X_2 affects X_3 . . . X_n affects X_1; the overall loop property can be designated either positive or negative depending on the number of negative linkages; the number of linkages in which a rise of the causative variable results in a drop of the affected variable, as opposed to positive linkages where causative and affected variables rise or fall together. A loop is positive feedback if the number of negative linkages is even, and negative feedback if the negative linkage total is odd. Positive loops tend to grow or decline without limit. Negative loops tend to restore to an equilibrium state. All the example systems at the start of Chapter 1 are negative feedback. The pendulum can be represented by a loop bob distance from vertical $\xrightarrow{+}$ force on bob $\xrightarrow{+}$ movement of bob $\xrightarrow{-}$ bob distance from vertical. This has an odd number (one) of negative linkages.

In the commodity system there are two negative feedback loops, chains of cause–effect tending to restore equilibrium. When stocks are lower than desired, there are price increases. These increases effect changes in both the production and consumption sectors which act to return inventory to its desired level. The price increases lower *per capita* consumption of the material. With less consumption drain inventories will tend to rise. Production relationships act with the same effect. The increase in current price raises producers' price forecasts and thus the capacity they wish to utilize. After acquisition delays, additions to capacity result in increased production and higher inventory. Were inventory to rise above its desired level, the resulting price decrease would again be propagated around both the production and consumption loops to yield counteracting forces, decreased production and increased consumption.

This system is not readily amenable to algebraic analysis because there are several processing delays to be considered:

1 Producing commodity;
2 Forecasting producers' price;
3 Forecasting consumption;
4 Recognizing retail price;
5 Responding to price forecasts;
6 Transferring productive capacity.

A computer simulation model can be constructed incorporating delays chosen to be of the right order for the commodity under study. Meadows investigated the pig cycle in the U.S.A. between 1925 and 1968 (an even longer series exists back to 1890). He was able to show that the most important delay in the system derives from the gestation and growth time of pigs, about 11 months.

Figure 2.3 shows the resulting runs from his model of the price and level of breeding stock over time: the cycle length is 45 months. The time series of average annual farm price is shown in Figure 2.4 and average cycle length here is 48 months, which agrees well in view of the 'noisiness' of the series. All factors external to the system (wars, depressions, outbreaks of swine-fever, etc.) are ignored. This is a necessary feature of all modelling exercises: a deliberate selection is made of variables to be incorporated. It is not necessary to assume that those variables omitted have no effect on the system, only that they are small in comparison with the effects of the selected variables, and random in their manifestation. The

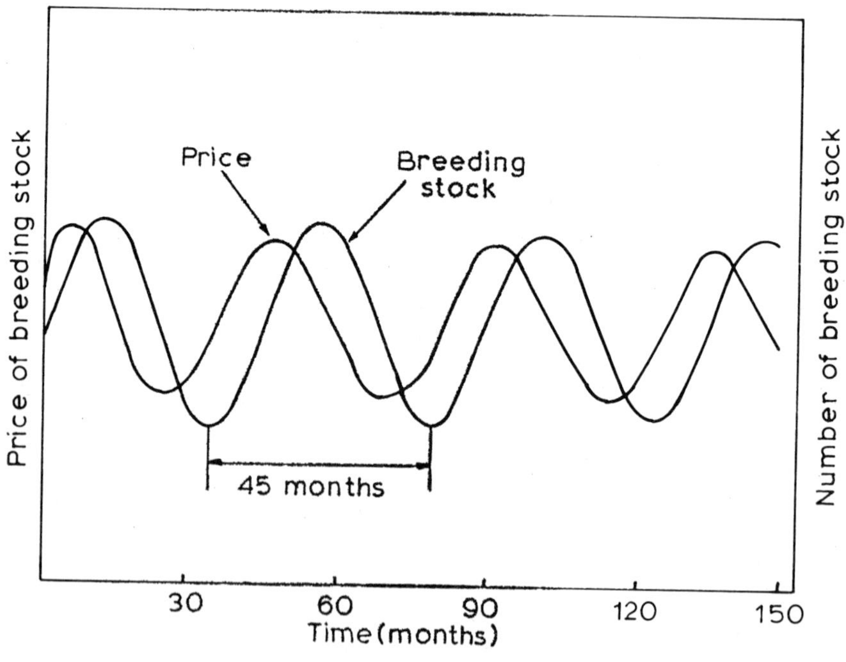

FIGURE 2.3 The resulting run from D. L. Meadows' investigation of the pig production cycle in the U.S.A. during the period 1925–1968. This shows a cycle length of 45 months in the rise and fall in breeding stock numbers, following the rise and fall in prices.

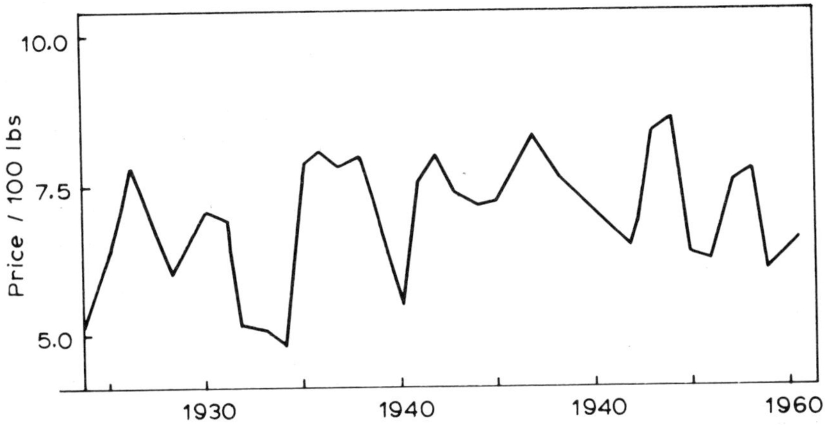

FIGURE 2.4 Time series of average annual farm price of pigs.

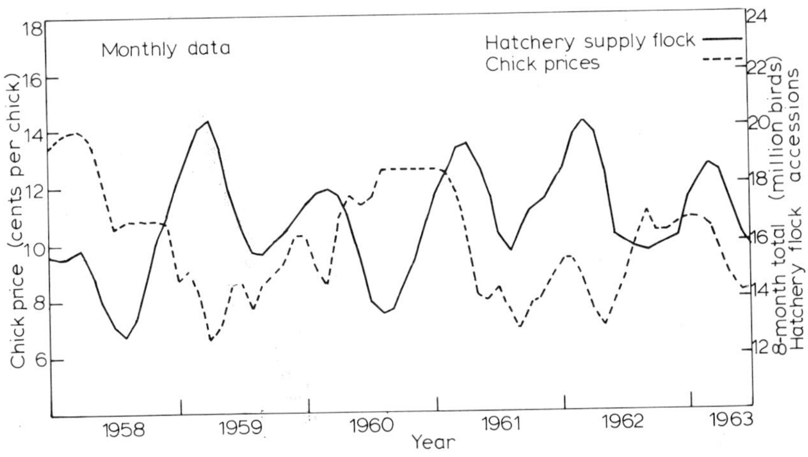

FIGURE 2.5 The chicken production cycle for the period 1958–1963 in the U.S.A. The model shows a period of 30 months compared with the observed period of 31 months.

random disturbances are often called 'noise' in analogy with the random disturbances present in communication systems, which manifesting auditorily are perceived as noise.

The Meadows hog-cycle model is particularly interesting because its application to other similar livestock commodities could be expected to reveal different cycle lengths for different production delays, and indeed this is so. For the chicken cycle, the time elapsing between the shipment of primary setting eggs to the hatchery and delivery of the first broiler eggs is about six months: time required to raise broilers from eggs is about three months. Other delays in the system are somewhat shorter than for the pig case because smaller risks are associated with investment in chickens. The model exhibits a period of 30 months compared with 31 months for the observed system. At the other extreme, cattle take a longer time for gestation and growth, the cattle cycle showing a period of 15 years.

Figures 2.5 and 2.6 show the cyclical behaviour associated with chicken and cattle respectively.

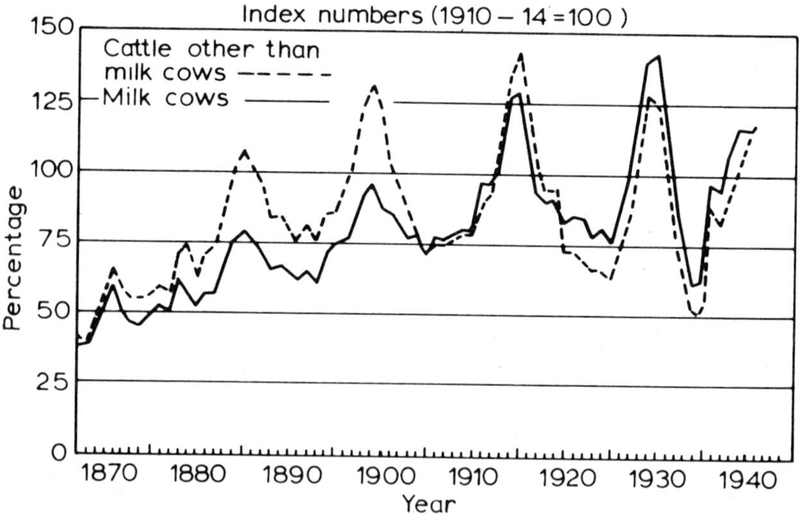

FIGURE 2.6 The cattle production cycle for the period 1870–1940 in the U.S.A. This shows a period of 15 years due to longer gestation and growth times (U.S. Department of Agriculture).

Models and thinking

Only a few examples have been offered here of modelling methods but the fields of application have been chosen to suggest a very wide scope for this style of thinking; a style firmly grounded in the physical sciences, both from construction, using simple elements, and from confirmation procedure requiring a good match to observable phenomena. It is appropriate at this point to probe more deeply into our reasons for preferring one model to another, and what is meant by declaring that a given model is 'true'. We all quickly acquire mental models of the way the world works, and these serve very well for our everyday activities. To ask why we believe in the soundness of these mental pictures is necessary for the proper appreciation of the significance of a scientific model, so the next chapter necessarily contains short philosophical excursions.

3. Prediction

The quality of a model is intimately associated with its power of prediction. A good model is one that enables predictions of the state of the system to be made with great accuracy. The further into the future that this accuracy is maintained, the better the model. For models with applicability to a diversity of systems, the power of prediction manifested in many different circumstances, enhances confidence in the faithfulness of the mapping between model and system.

The statements made above are necessarily qualitative because it is only at the local and specific level that quantitative comparisons are useful. Thus a measurable comparison can be made between the Ptolemaic and Newtonian models of planetary motion, and it can be said that after a given number of years, the error in the predicted position of Mars is so many times greater using the Ptolemaic model than that resulting from the Newtonian model. Such a comparison is not helpful if used in relation to different models of different systems. The accuracy of prediction of a weather forecast is certainly lower than that of a solar eclipse forecast. Nevertheless, whereas in the first example the comparison leads to discrimination and the discarding of Ptolemy in preference to Newton, one cannot say that eclipse forecasting is preferred to weather forecasting. The appropriate comparison would be based on the intrinsic predictability of the two systems. The compiler of the nautical almanac can justly be accused of gross error if he shows an occultation separated by 10 minutes from its actual time of occurrence. The weather forecaster is to be congratulated if he names the first fall of winter snow to within a day.

Noise

It may be objected that to introduce the term 'intrinsic predictability' is to invoke an absolute where none exists. Forecasting performance could be merely local to time and place like speed performance. In the early years of aviation a speed of 200 miles an hour was remarkable, but now supersonic flight is commonplace. However, there are good reasons for believing that the idea of an absolute limit to predictability is not only sound in concept but that under certain conditions the limit is itself ascertainable.

In communications engineering 'noise' is an ever present reality. The variations in electric current that carry the information in the circuits all take place against an unending background hiss. When the disturbances from faulty components and poor contacts have been eliminated, there remain the underlying random thermal motions within the conductors which is called 'noise'. Free from conductors, the signal carried on electromagnetic waves, still contends with background waves of random amplitude and phase. Apart from transmission through wires, waveguides or free space, the transducers at all stages contribute their own noise. The notion of a pure signal is foreign to the communications engineer: the signal-to-noise-ratio is his reference measure and though this ratio may be high it is never infinite. In communications there is an agent responsible for transmission and the object of the engineer is to present the agent's message with as little corruption as possible, that is with the least noise addition. In scientific work it is assumed that there are patterns or messages in the wealth of impulses reaching us from the physical world, and that by suitable techniques the messages can be revealed in a pure enough form for us to be convinced of their reality. In Figure 3.1(a) is shown a simple sine wave corrupted by successively larger noise components, random fluctuations superimposed on the sine wave. With small noise there is little doubt of the reality of the underlying wave, but at the last stage illustrated (3.1(e)), there is little reason to guess at the presence of the wave.

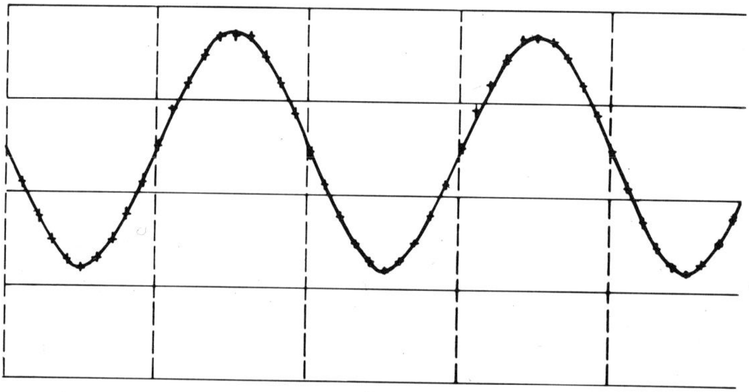

```
                 AMPLITUDE
PERIOD 0          0.2         0.4         0.6
 60.00 .
 30.00 .
 20.00 ***************************
 15.00 .
 12.00 .
 10.00 .
  8.57 .
  7.50 .
  6.67 .
  6.00 .
  5.45 .
  5.00 .
  4.62 .
  4.29 .
  4.00 .
  3.75 .
  3.53 .
  3.33 .
  3.16 .
  3.00 .
  2.86 .
  2.73 .
  2.61 .
  2.50 .
  2.40 .
  2.31 .
  2.22 .
  2.14 .
  2.07 .
  2.00 .
  1.94 .
```

FIGURE 3.1 (a)

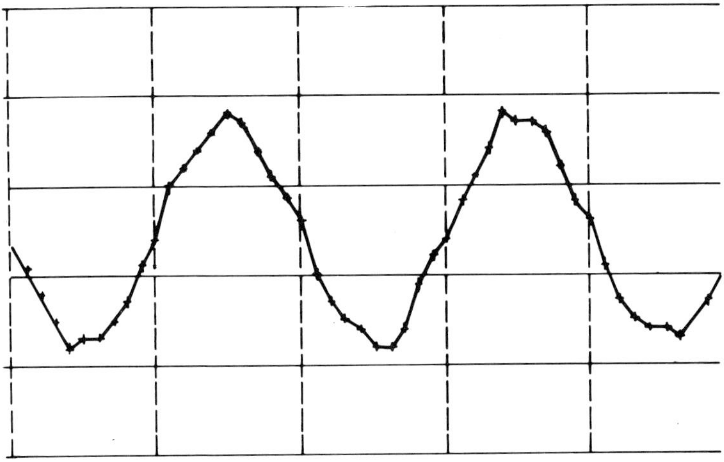

```
                 AMPLITUDE
PERIOD 0            0.2          0.4          0.6
 60.00 .
 30.00 .
 20.00 **************************
 15.00 .
 12.00 .
 10.00 .
  8.57 .
  7.50 **
  6.67 .
  6.00 **
  5.45 **
  5.00 .
  4.62 **
  4.29 **
  4.00 .
  3.75 .
  3.53 .
  3.33 .
  3.16 .
  3.00 .
  2.86 .
  2.73 .
  2.61 .
  2.50 .
  2.40 .
  2.31 .
  2.22 .
  2.14 .
  2.07 **
  2.00 .
  1.94 **
```

FIGURE 3.1 (b)

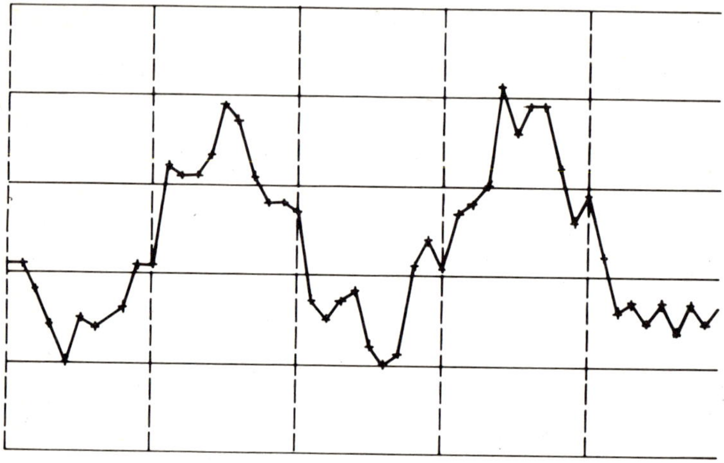

```
          AMPLITUDE
PERIOD 0           0.2          0.4          0.6
 60.00 ***
 30.00 **
 20.00 ************************
 15.00 **
 12.00 .
 10.00 **
  8.57 **
  7.50 ***
  6.67 **
  6.00 ***
  5.45 ***
  5.00 **
  4.62 ***
  4.29 ***
  4.00 ***
  3.75 .
  3.53 ***
  3.33 **
  3.16 ***
  3.00 **
  2.86 ***
  2.73 **
  2.61 **
  2.50 **
  2.40 **
  2.31 ***
  2.22 .
  2.14 .
  2.07 ***
  2.00 **
  1.94 ***
```

FIGURE 3.1 (c)

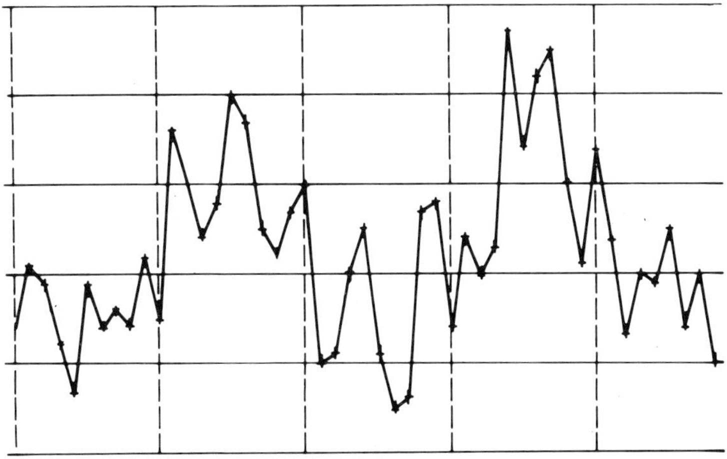

```
          AMPLITUDE
PERIOD 0              0.2          0.4          0.6
 60.00 *****
 30.00 ***
 20.00 *********************
 15.00 ****
 12.00 **
 10.00 ****
  8.57 ****
  7.50 ******
  6.67 ****
  6.00 ******
  5.45 ******
  5.00 ***
  4.62 *******
  4.29 *******
  4.00 ******
  3.75 **
  3.53 *****
  3.33 ***
  3.16 ******
  3.00 ****
  2.86 ******
  2.73 ****
  2.61 ***
  2.50 ***
  2.40 ****
  2.31 *****
  2.22 **
  2.14 **
  2.07 *******
  2.00 ***
  1.94 *******
```

FIGURE 3.1 (d)

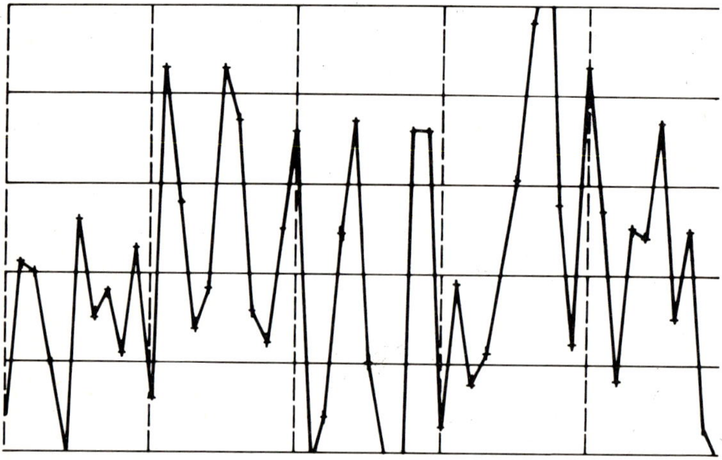

```
            AMPLITUDE
PERIOD 0            0.2          0.4          0.6
 60.00 ********
 30.00 *****
 20.00 *****************
 15.00 ******
 12.00 ***
 10.00 ******
  8.57 *******
  7.50 ***********
  6.67 ******
  6.00 **********
  5.45 ***********
  5.00 *****
  4.62 ************
  4.29 ************
  4.00 **********
  3.75 **
  3.53 ********
  3.33 ****
  3.16 **********
  3.00 ******
  2.86 **********
  2.73 *******
  2.61 ****
  2.50 ****
  2.40 ******
  2.31 *********
  2.22 **
  2.14 ***
  2.07 ************
  2.00 *****
  1.94 ************
```

FIGURE 3.1 (a–e) Steps in the corruption of a simple sine wave shown in *a*, by the gradual addition of 'noise' to the signal. The actual frequency content for each graph is shown beneath it.

FIGURE 3.1 (e)

The diagrams of Figures 3.1(a) to (e) have been produced by synthesis: adding noise to the signal. The actual frequency content is shown below in each case. 'Period' refers to the interval between successive peaks or troughs of a wave, and 'Amplitude' is the ratio of excursion from the mean to mean value. Random fluctuations have frequency components of equal amplitude throughout the frequency spectrum, so the signal component appears as submerged more and more completely into this uniform amplitude level. The reverse process to that of synthesis is analysis: the teasing out of the signal from the mixture as presented. In Figure 3.2 is given the frequency components of the U.S.A. hog price data of Figure 2.4. The Meadows model is an attempt to compose a structure delivering an output corresponding to the chief component of this spectral analysis. The discrepancy between the chief component identified in the spectral analysis (period of 5·29 years) and Meadows' observed figure of 4 years arises because he took the period of the most recent complete cycle.

Models as predictors suffer from noise in two ways. First, the determination of the parameters of the model is performed in the

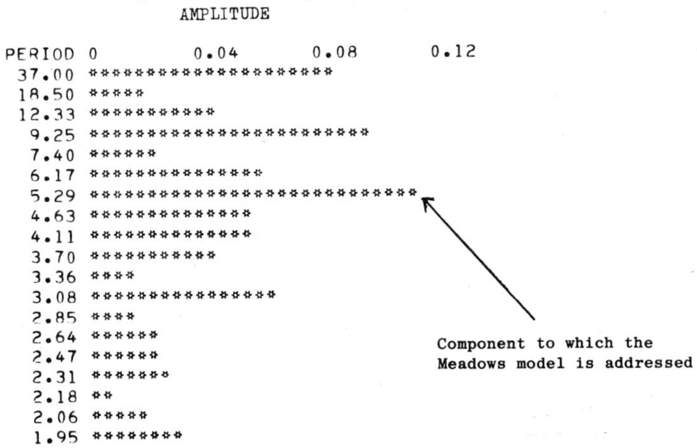

FIGURE 3.2 The U.S.A. pig price data shown in Figure 2.4 is here translated into frequency components.

presence of noise. In the case of the Meadows model, the two parameters are the amplitude and phase of the chief spectral component frequency. Secondly, the time period in which the prediction is made is subject to noise, so that even if the parameters were known with complete accuracy the prediction would still suffer from the continuing presence of noise. In the physical sciences it is more customary to speak of the error associated with observations, parameter determinations and inferences. The terms 'noise' and 'error' are not quite synonymous because error is further partitioned to 'systematic errors' (bias of some kind) and random error; only the second of these two can be interchanged with noise. Errors of measurement are quite inevitable. An observation of perfect accuracy would imply an infinite amount of information, an infinite number of bits: this would require infinite energy outlay.

Redundancy

Even when the signal-to-noise-ratio is small, complete transmission of the message can take place by using redundancy. In its simplest form this entails repeating the message again and again. Provided that sufficient repetitions are permitted, the signal can be as weak as we choose, the transmission consisting almost entirely of noise. Table 3.1 shows a binary transmission of the word CAT by using groups of 5 digits, each to represent one letter according to its numerical position in the alphabet. Thus 'C' is the third letter and is represented by 00011 ($2 + 1 = 3{:}\,00010 + 00001 = 00011$). In the same way, 'A' is represented by 00001 and 'T' by 10100. Line 1 is the message to be sent. Lines 2 to 6 are repeated transmissions where about 10% of all the digits are corrupted (0 becomes 1 and 1 becomes 0). None of these 5 lines are correct, but in line 7 the 'majority verdict' has been taken for each of the 15 positions of the 5 repeats and the result is again 'CAT'.
The same principle is used in making observations. By repeating the observation many times and averaging the results, the effect of random error is reduced. The reason for the error reduction lies in the fact that random effects are uncorrelated: they are as likely to

TABLE 3.1 Binary transmission of the word CAT by redundancy method

	5 digit codes for letters	Word transmitted
Original transmissions	0 0 0 1 1 0 0 0 0 1 1 0 1 0 0	CAT
Repeat transmissions	0 0 0 0 1 0 0 0 0 1 1 0 1 0 1	AAU
	0 1 0 1 1 0 0 0 0 1 1 0 1 0 0	KAT
	0 0 0 1 0 1 0 0 0 1 1 0 1 0 0	BQT
	0 0 0 1 1 0 0 0 0 1 1 0 1 1 0	CAV
	0 0 0 1 1 1 0 1 0 1 1 0 1 0 0	CUT
Majority	0 0 0 1 1 0 0 0 0 1 1 0 1 0 0	CAT

increase the magnitude of the observation as to decrease it. In a simple experiment to measure the probability of a tossed coin showing heads, the absolute excess (or deficit) of heads over tails will generally increase as the number of tosses is increased, but the excess (or deficit) per toss will tend to zero, and the probability of a head tend to 0·5, for an unbiassed coin.

The repeated message and the repeated observations are examples of serial redundancy. But parallel redundancy can be used to achieve the same effect. For example, the message CAT could be sent simultaneously by 5 separate transmitters and the majority verdict used to overcome the transmission defects. In the case of the coin, many coins could be tossed simultaneously (rather than tossing the same coin repeatedly) and the proportion of heads interpreted as the probability. (It is arguable that Gregor Mendel would not have discovered the laws governing inheritance of characteristics if he had been obliged to rely on serial rather than parallel redundancy and cultivate only one pea at a time.)

Consider these concepts of noise and redundancy in relation to one previous example system, the solar system. The problem of predicting the future position of a particle of dust in a cloud occupying interplanetary space is formidable. Impacts from other particles coming from all directions will disturb its trajectory in a random fashion.

Radiation pressure from the sun will influence its movement significantly: not in a uniform fashion because the radiation will be scattered and reflected by the other particles present. It is apparent that the problem would become much easier if we wanted to predict the position of the centroid of the cloud: and even easier if the cloud of particles coalesces into a solid lump. Indeed the bigger the lump the proportionately smaller are the disturbances from particle impacts and radiation pressure. The large lump constitutes massive parallel redundancy. The random movements relative to one another of all the particles comprising the lump have a net effect on the movement of the lump's centroid, which becomes less and less significant the larger the lump size. A hurricane raging on some area of the earth's surface produces no change in the orbit that we can measure. Some two thousand tons of assorted debris strikes the earth every day at speeds greater than seven miles a second with no measurable orbit consequence.

Because the noise with which we have to contend in predicting planetary motion is relatively very small our forecasting performance is impressive, but it should be noted that it is limited in two distinct ways. The parameters of the system are limited in the accuracy to which we know them. This could be improved by devising better measuring apparatus and making more observations, but ultimately we come up against the natural limit from the noise which is present, and this source of error is cumulative. In Figure 3.3 is shown a predicted trajectory for an object as a straight line, and the actual path of the object as a disjointed line because of successive random discrete disturbances. Each disturbance comes after a train of such prior effects and adds to the total error. If the errors were successively $+2$, -1, 0, $+1$, the net error would be $+2$. Although the errors are assorted positive and negative they do not, in general, total zero overall and the effect is a drift away from the calculated trajectory. If in each time interval t a transverse error drawn from a population of errors with standard deviation s affects the trajectory, then it can be shown that after a time nt the standard deviation of the expected cumulative error will be $s\sqrt{n}$.

Inaccurate system parameters result in a consistently positive or negative error prediction which increases with time lapse, and

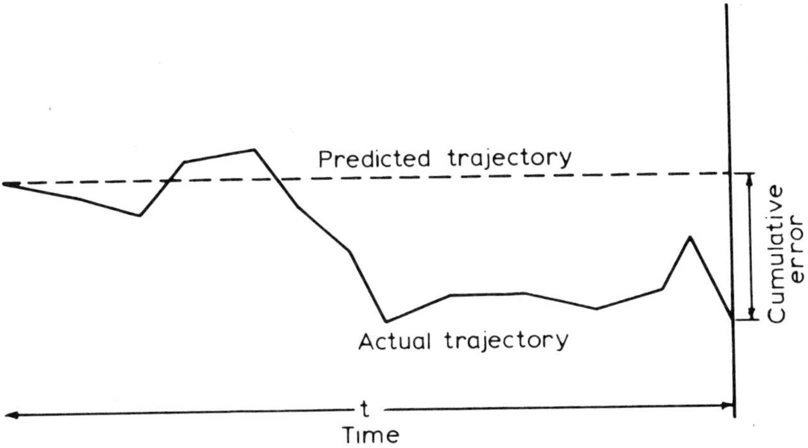

FIGURE 3.3 The diagram shows the effects of successive, random discrete disturbances on a predicted trajectory.

ambient noise produces a similarly increasing error but of no fixed sign. Even in our most predictable system these error sources are present. Solar eclipses can be retrodicted a thousand or even two thousand years ago to the day and the hour, but not a million years ago; and likewise predicted a thousand, but not a million, years ahead.

Prediction efficiency

Because there exists a very wide spectrum of size for the objects present in the solar system, from molecules of gas at 10^{-27} kg to the sun at 10^{30} kg, the progression of parallel redundancy is readily appreciated; the concept of intrinsic predictability attains a meaning which is independent of our ability to devise efficient predictors. The concept has further utility in that efficiency of prediction is more appropriately measured against a standard specific to the system than in absolute terms. A parallel from thermodynamics may help here. Consider a heat engine designed to take in heat energy Q from a source at temperature T_1 to deliver work W, and to reject heat to a sink at temperature T_2. There are two separate efficiencies for the engine. The first and most obvious is:

$$\text{Efficiency} = \frac{\text{work delivered}}{\text{heat energy from source}} = \frac{W}{Q}$$

However, it is known from the second law of thermodynamics that the most efficient engine conceivable (not merely the best one so far devised, or even the best one that ever will be devised) has an efficiency:

$$\frac{W}{Q} = \frac{T_1 - T_2}{T_1}$$

Thus, the intrinsic efficiency potential for a source–sink combination depends on the two temperatures involved. Elementary calculation shows that there is a truly immense amount of solar energy absorbed by the upper layer of the ocean. The actual amount of recoverable work by using the upper layer as source and the deep layer as sink is more modest. Suppose the upper layer is at 27°C (300°K) and the deep layer at 7°C (280°K). The efficiency of the best conceivable engine is:

$$\frac{W}{Q} = \frac{300-280}{300} = 6{\cdot}7\%$$

Contrast this with a solar generator designed to operate by concentrating the sun's rays on to its receiving component. The temperature of the sun's surface is approximately 6000°C and the sink for the generator (assumed mounted terrestrially) about 20°C.

$$\frac{W}{Q} = \frac{6273-293}{6273} \simeq 95\%$$

These figures correspond to what have been called first law efficiencies: they are simple ratios between input and output energies. An alternative efficiency is the second law efficiency, which is the ratio of the work done by the real heat engine to that done by an ideal heat engine, working between the same temperatures and using the same quantity of input source heat. As examples of the use of the two efficiencies, suppose a modern successor to Claude (the first engineer to attempt using marine temperature differences for generating power) who obtains a first law efficiency of 5% with temperatures of 27°C and 7°C.

$$\text{Second law efficiency} = \frac{5}{6{\cdot}7} = 75\%$$

The best solar generators at present reach first law efficiencies of about 40%:

$$\text{Second law efficiency} = \frac{40}{95} = 42\%$$

Our modern Claude, though he converts a meagre 5% of the solar energy trapped in the sea, is clearly a very good engineer.

A model is conveniently regarded as a prediction machine and the performance of the machine is to be judged against the best conceivable machine. The analogy can be carried further. Improvements in heat engines have been characterized by step changes: mechanisms which are different in kind. The reciprocating steam engine, the steam turbine, the magneto hydrodynamic generator are quite dissimilar devices. Improvements in prediction through better models result typically from structures which are radically different from their predecessors. The Newtonian model is not an incremental improvement on the Ptolemaic. The Rutherford atom with all positive charge in the nucleus is quite unlike the Thomson 'plum pudding' atom with distributed positive and negative charges.

The ideal heat engine and the ideal prediction machine are linked precisely if ambient noise is paired with the entropy of the source–sink combination. If the temperature difference between source and sink is zero, the entropy is a maximum, that is the disorder present is a maximum: no work can be extracted. This corresponds in the prediction case to complete domination by noise, and the model provides no reduction in error for any prediction made (the prediction is meaningless). An ideal heat engine is one which produces the maximum work output possible for a given level of disorder (given temperature difference). An ideal predictor is one which generates predictions as accurate as they could be for the given level of noise. Thus the ideal predictor is one which has the best structure and absolutely accurate parameters. If one is prepared to accept bestness of structure, the 'second law efficiency' of the predictor has a strict analogue in the accuracy of the parameters. One flaw prevents the completeness of the pairing. Though we can know the limit to the heat engine's efficiency without reservation, the predictor may not have the best structure and so the accuracy of its parameters gives an optimistic estimate of its efficiency: its ability to generate accurate predictions allowing for ambient noise. An example of this appears in the Newtonian model. It is not quite correct because it needs slight modification from relativity effects. Relativity effects obtrude when velocities are high and the orbit of Mercury (the planet with highest velocity) is more accurately predicted if Newton is modified by Einstein.

Now the fact that a later correction to the Newtonian model has appeared suggests that further refinements may be possible and this uncertainty brings out the difference between heat engine and predictor. For the heat engine, the ideal is calculable because the quantity of latent disorder in the source–sink combination is known precisely from their temperatures. To perform the same feat for the predictor would entail knowing the latent level of randomness present in the system to which the predictor is applied. In particular cases we may have shrewd suspicions, but no means of being sure. For example, biologists doubt the possibility of improving much on the fit to Nicholson's sheep blowfly data given in Chapter 2, but it would not be unreasonable to suppose that the Meadows hog-cycle model could be improved, by taking into account linkages to the rest of the economy.

Plausibility

A further parallel between model and machine can be traced. The machine is constructed from a set of components each of which is simpler than the machine. A simple component is one manifesting less variety of behaviour than the full functioning machine. In the same way, a model is assembled from simpler parts. The reason that an explanation of a phenomenon in terms of a model is acceptable derives from familiarity with the properties of the components: those properties are necessarily simple. Imagine yourself explaining the occurrence of a lunar eclipse to a child. You could use a tennis ball, a football and an unshaded electric light bulb to represent moon, earth and sun as the components of your model. As you bring the football between the bulb and the tennis ball so that the shadow, with curved edge, falls on the tennis ball the process seems to be 'explained'. The point is that tennis balls, footballs and electric light bulbs are familiar objects to the child and shadows are familiar occurrences. If more elaborate pieces of apparatus are employed the explanation is less convincing: any unfamiliar object is immediately suspected of having novel properties and itself contributing to the result. (This is an entirely appropriate response with good survival value; watch a kitten approaching a tortoise for the first time.)

The idea of familiarity—and because familiar also plausible—is a pointer to the distinction between devising and testing models. Models and machines are built from familiar components. There is a progression in what we find familiar. An amplifier is built from resistors, capacitors, transistors and connectors. In this list of components, the transistors have been familiar for only the last 15 years. Thus familiarity is rather like fashion: triode valves are becoming unusual in the experience of this generation of electronic engineers and possibly electric light bulbs will become obsolete and replaced by discharge tubes and light emitting diodes. And the plausibility that derives from familiarity can be deceptive. All those changes in fashion for the way in which children should be reared were doubtless appealing at the time that each was in vogue.

Plausibility is necessary for the model to be envisaged at all. Plausibility is necessary for it to be received sympathetically by others. But it is not sufficient. The sufficiency lies in the testing: the presence of latent ability in the structure to predict. Compare the case of the machine and the drawings/instructions that specify its construction. The parts are obtainable, the instructions are clear, the means of assembly is evident, the account of the machine's operation is understandable: but when the construction is complete, the machine does not actually work. Examples of this can be found among the patents filed by inventors who have been plausible enough to convince themselves and the Examiner but who fail the acid test. It does not matter if inventors try and fail, but models, simple or complex, sound or unsound are used everyday as the basis for decisions. The late Sir Malcolm Campbell held the view that the speed limit for vehicles in cities should be raised to at least 40 m.p.h. He reasoned that at higher speeds the driver's concentration increased and that by virtue of the greater attention to hazards, the accident rate would be lessened. The first part of this argument can be confirmed by personal experience. One of the attractions of pastimes like ski-ing and tobogganing is the intense concentration needed to negotiate a track at high speed. However, the last part of the argument fails the touch-stone of experiment. There is a relation between speed and accident rate measured by fatalities or injuries, which shows them increasing together over a wide range of speed. Trials in different countries confirm this relation. Campbell's model of driver, vehicle, speed, concentration level and accident rate is defective. Even the football, tennis ball and bulb model of a lunar eclipse is only a beginning, and the percipient child will demand more justification before you convince him.

The distinction between devising and testing is worth examining at some length. Inventors choose those components readily to hand, and the functions of the devices which they construct reflect their interests. The same holds true for the models of science. Scientists have constructed models of those phenomena which interest them and their compositions are made from the equipment available at the time: often the mathematical contrivances of the century before them. The extensive development of statistics owed much to the analysis of probability theory which, in turn, flowered as a result of the popularity of gambling and the fascination of mathematicians with the paradoxes of gambling. Connections between the culture of a society and its scientific models abound; so it was not necessarily pejorative for the Soviets to label Mendelian genetics as bad Marxism, in comparison with the politically more congenial Lysenko model which emphasized nurture rather than nature. Once again matters had to be decided by empirical tests and eventually the Lysenko acclimatization regime followed Lamarckian inheritance, phlogiston, caloric, etc. into the dustbin of discarded theories.

The manner of reception which a theory or model receives at its first announcement is a fascinating area of study for the historian of science because it is so revealing of the often submerged, concealed primal convictions of people. World modelling initiated by Jay Forrester in 1971 evokes similar reactions to those which were provoked in their time by the notion of a heliocentric universe and by the Darwinian theory of evolution. All three strike at the conviction of man as a unique, free agent. First he is dethroned from the centre of the universe, then he is assigned a place (and a very recent one) in the evolutionary progression of animals, and now his breeding and his appetites are to be constrained and fettered by the limits of his planet. More interesting has been the form of criticism directed at world modellers. It is said that these models reflect only the prejudices and foregone conclusions of the modellers. The inference intended to be drawn from this comment is that the models need not therefore be taken seriously. The truth is that all models embody the preconceptions of the modeller: indeed a prevision of the machine or the model is essential to the designer. It would be absurd to ignore a new machine because the inventor held alien political beliefs. The apposite stance is inquiry as to whether it works. It is so with world models. Not 'Is he a doomster/optimist?' but 'How well does his model mirror the real world?'

Past and future

Given that prediction is key to modelling, the meaning of tense must be examined. For laboratory systems the meaning is clear, the model can be used to predict the result of a future experiment. For a system where experiment is impossible, difficult or undesirable, in astronomy, meteorology and economics respectively, the point in time called 'present' is less definite. Suppose that the modeller receives the data on values of the variables in the system from a monitoring centre. His interest is in making predictions a month ahead and he received the figures from the monitoring centre at weekly intervals. Thus he is able to check the accuracy of his predictions every month. Then the monitoring centre starts to slip behind, sending out the figures later and later so that eventually the modeller is receiving in March the data which applied to the system state in January. However, the sets of figures are still in sequence and his regime is quite unperturbed. He continues to check predictions made a month earlier against figures arriving in the current week. The method is as much valid now as when the centre was punctual, except that the 'predictions' are no longer statements about the future because at the time they are made the system has already passed that point. So perhaps prediction is an unnecessarily strong requirement and only consistency between different parts of the data series is needed? If consistency between the data sets for January–August and that for September is acceptable, why not between February–September and January, or even January–April plus June–September and May?

The words 'needed' and 'requirement' above imply a new question: 'Does the consistency found by the use of the model establish the truth of the model?' This may seem a harmless enough question, but it raises very severe difficulties. It is a common experience for modellers to discover that a very simple structure is adequate for short term forecasting but that only a more elaborate structure suffices for the longer term. In some cases the simple model is superior in the short term in that it yields smaller errors in the predicted values of specified variables than are obtained from the more complex model. By what criterion are we to decide which of the models is more true? Clearly the word 'true' is difficult to define in these circumstances. To return to the comparison with the machine, there can be machines intended for the same purpose but whose ranking is reversed by a change of environment. A light weight racing bicycle is superior to a rugged one when used on a smooth track but not over bridle-paths. Language would become stilted and tedious if words like 'truth' were proscribed, but for this discussion the criteria will be made explicit.

Consistency

The role of consistency can be put shortly. A model capable of predictions having small associated errors will be found to have high consistency in the data series interpreted through the model. The converse statement cannot be made. An example to illustrate this point is offered below.

t	0	1	2	3	4	5	6	7
x	50	38	63	74	70	28	57	36

This series could have been generated from the summed output of 4 oscillators:

$$x = 52 + 8 \sin (\pi t + 1{\cdot}6) + 7 \sin (3\pi t/4) + 11 \sin (\pi t/2) + 14 \sin (\pi t/4 - 0{\cdot}8)$$

All eight values are consistent with the expression above. The next value of x (for which $t = 8$) is deduced from the formula to be 50. In fact the next value was 41. We are all immediately suspicious of a prediction formula that contains as many, or nearly as many, parameters as there are items in the input series, for the simple reason that given the freedom to use a number of parameters equal to the number of items, an expression can always be fitted. Indeed, more than one expression can be fitted: a polynomial could have been used above with the coefficients being the parameters rather than the amplitudes and phases of the sine waves.

The models of the physical sciences that are called 'laws' are notable for the very small number of parameters required to achieve consistency and to make predictions. Given a set of known masses and the separating distances, all of the interattractive forces can be predicted using only one parameter, the gravitational constant G. So perhaps there can be a substitute for prediction if a model satisfies both the requirement of consistency among the input data and the requirement to perform this feat with a number of parameters much smaller than the number of input items of data. Consider that relationship which still receives honourable mention in the text

TABLE 3.2 Comparison of actual and calculated atomic weights of common elements

Element	Specific heat	Predicted atomic weight	Actual atomic weight	% error
sodium	0·264	22·8	23·0	− 1
aluminium	0·203	29·6	27·0	+10
calcium	0·149	40·3	40·1	+ 0·5
iron	0·105	57·2	55·8	+ 3
copper	0·089	67·5	63·6	+ 6
zinc	0·088	68·3	65·4	+ 5
silver	0·053	113	107·9	+ 5
cadmium	0·052	115	112·4	+ 3
iodine	0·050	120	126·9	− 8
mercury	0·029	207	200·6	+ 3
thallium	0·031	194	204·4	− 5
lead	0·029	207	207·2	− 0-1

TABLE 3.3 Further comparisons of actual and calculated atomic weights

Element	Specific heat	Predicted atomic weight	Actual atomic weight	% error
boron	0·304	19·8	11·0	+ 80
carbon	0·113	53	12·0	+340
silicon	0·179	33·5	28·3	+ 18

books as the law of Dulong and Petit. One way of expressing it is to say that you can predict the atomic weight of an element if you divide its specific heat into 6. Table 3.2 gives some of the common elements to show the law in action.

Is your confidence growing? (Only one parameter is required.) In Table 3.3 are some more elements.

Although the second list (Table 3.3) is a disappointment, it must be more than coincidence that the predictor worked so well in the first list (Table 3.2). The fuller knowledge of vibrations, within the atomic lattice, that came after the time of Dulong and Petit revealed why the predictor is good for some elements and poor for others. Their 'law' must be labelled a local relationship. If it had worked only twice we should have been inclined to dismiss it as a coincidence. This brings out a feature of models which is easily sensed but difficult to formalize. Some physical laws are regarded as more secure than others. A group of scientists invited to rank laws in order, starting with the ones for which they have most doubt that they will be disproved, produce similar lists. The property of 'Security' is not to be had by mere repetition. The scepticism of David Hume about induction has survived attempts to put it on a firmer footing.

If a litre of sodium carbonate solution is added to a litre of calcium chloride solution, a voluminous white precipitate is produced. One can divide the solutions into 1000 parts each of one millilitre and then mix the parts in a 1000 separate operations so that two millilitres of mixture result each time, and the precipitate appears each time. Compare this laborious and uninspiring experiment with three others in which first, calcium nitrate solution is substituted for calcium chloride solution, second, potassium carbonate is substituted for sodium carbonate and third, both of these substitutions are made. In every case the white precipitate appears and by suitable tests can be shown to be chemically identical in all four instances. The essence of establishing a scientific law is to test it under a great variety of circumstances: to arrange for widely differing values to be taken by all the variables which could possibly affect the relationship under study. In this process, a relationship which starts by being only local may advance to be regarded as a general universal law. To the question of how we know which variables could possibly affect the relationship, there is no straightforward answer. An investigator might well repeat his experiment with the laboratory in darkness, but he will be unlikely to worry if the light in the next room is on or off. A matrix display with phenomena marking the rows and variables marking the columns and having entries wherever an effect of one on the other is known to occur is, for our particular terrestrial environment, sparse. If it were not so, if the investigator could not assume that practically everything going on outside his laboratory is irrelevant, then development of science would be nigh on impossible. Parts of the matrix were known or suspected in pre-scientific days, and every modern investigator takes advantage of the painstaking work of filling up more entries that has been going on for three hundred years. To try out a new relationship in the 'hard' sciences under many different conditions is easier now than it has ever been. In the soft sciences it is still difficult. Phillips[5] suggested a relationship in economics which was intuitively appealing and showed a strong correspondence (an inverse relation) between observed values of unemployment and the rate of change of money wage rates in the U.K. between 1861 and 1957. However, in recent years the Phillips curve has not been a good predictor. Possibly the underlying concept of the labour market resembling a commodity market is sound, but the coefficients in the relationship are affected by other factors in the economy, and change over time.

Recapitulation

1 It is necessary for a model to be internally consistent, to depend on only a few parameters and to be demonstrably capable of prediction.

2 Prediction is to be construed in a more general sense than using data from past observations to predict the values which future observations will take. Out of a set of data applying to the system being modelled, the model must predict, using information derived from only one part of the set, the values appearing in the other part. The absence of information from the subset to be predicted is crucial to the test.

3 Two models intended to perform the same function can be compared through the sizes of the errors generated in predicting a given variable at a given point in time.

4 All models are subject to noise. A model can be improved in the accuracy of its predictions only up to the point where the predictive error derives solely from the noise.

5 There is a second qualitative measure in addition to predictive accuracy and this is security. The more varied the conditions under which a model is tested successfully, the higher the security.

Measurement and testing

The discussion so far has been in terms of predicting point values because this is a concept which is readily understood, but the manner of measurement is open to wide choice. Usually the value predicted and the measurement made are averages of some kind. The 'position' of a celestial body refers to its centroid, which is a weighted average of its constituent particles. Sales volume measured over a month or a quarter amounts to averaging the daily sales. The scale used need not be linear. These are linearly based averages but hydrogen ion concentration is commonly measured on a pH scale which is logarithmic. The interest may centre not about the first moment but the second (the variance). Though minimum predictive error determines model choice, the guidelines for assembling model structures likely to be good predictors often stem from more recondite analysis. It will be recalled from Chapter 1 that one of the deductions to be made from the molecular model for a gas was the negative exponential distribution of molecular energies. Experimental determination of the actual distribution confirms the deduction as correct and this strengthens belief that the structure of the proposed molecular model is sound.

In most cases it is possible to prescribe tests which enable one model to be preferred to another in respect of some aspect of behaviour which is manifested. However, it may not be possible to state how closely the model approaches the best model that could be constructed. The residual error may be ascribed to noise. This means that the limit of predictability has been reached. However, the model structure could be incomplete and a further refinement permit smaller residual error. As a homely demonstration of regression methods to a class of managers, I used to take their heights and weights as the variables and construct the histogram of residual errors for weight predicted from height. Although the shape of this histogram usually approximated to normal, the prediction error was always reduced by using, as an additional variable, the waist measurement in the regression. This result is unsurprising, but it should be noted that the first essays in conjuring order out of apparent chaos are often greeted sceptically, indicating the general belief that no further improvement is possible. When the first paper was delivered outlining the regularities in the properties of elements according to an arrangement which we now call the 'periodic table', one sceptic asked the speaker if he had tried arranging the elements in alphabetical order of the initial letters of their names!

Security and truth

Under (5) of the summary it might appear that the degree of security can be interpreted as the probability of the model being true. Although there could be no objections to defining a statement as true because it is completely secure this supplies only an upper limit and can be used only about a special class of statements. For example the relation:

$$e^{4\pi i} = 1$$

is completely secure: it is true; but it is an identity and reveals nothing about relationships in the material world. Models of the real world must always be qualified by references to the particular conditions of the tests. If one wished to refer to a probability of being true it would be necessary to know about all other possible test conditions; clearly the testing can never be exhaustive nor can one know the extent to which exhaustive testing has been approached. Thus the context is incompletely specified and a probability cannot be assigned. This feature of models is not unique nor even particularly rare. If a chord is drawn at random in a circle it appears that the probability of one of the two arcs being at least $\frac{2}{3}$ of the circumference of the circle could be calculated. How will the randomness be defined? Three possible ways are illustrated. There could be equal probability associated with each degree of angle made with a tangent over the full 180° (Figure 3.4). The probability could be equal for each

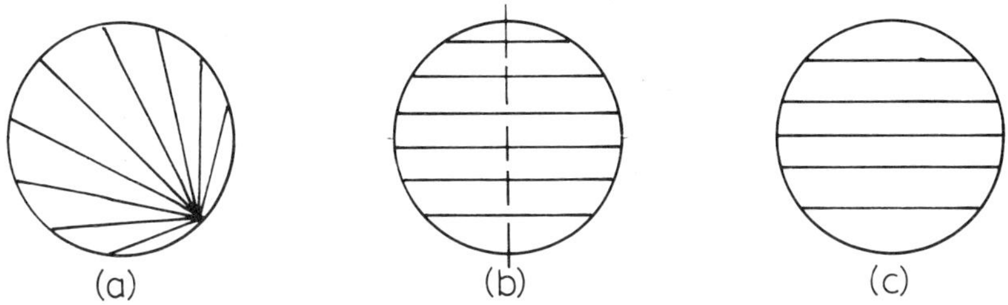

FIGURE 3.4 *a–c* Three possible ways of defining randomness. Chords drawn within a circle so that one of the two arcs formed will be at least $\frac{2}{3}$ of the circumference of the circle. (a) equal degree increments; (b) equal intervals of distance across a diameter cut at right angles; (c) equal increments of area.

interval of distance across a diameter through which possible chords pass at right angles (Figure 3.4b). Equal increments of area cut off by the chord could correspond to equal probabilities (Figure 3.4c). These three ways yield three different answers to the probability question posed. The context is incompletely specified and the answer is indeterminate.

A model is made more secure by an additional successful test under novel circumstances. How secure it is compared with another model of another system is not known. The reason why scientists easily reach consensus on which models should be considered the most secure is partly because of the length of time that those basic models have stood (and therefore the number of opportunities for disproof they have survived), and partly because of the interweaving and overlapping character of the basic models: the whole fabric would be in danger if one of them became suspect. However, the occurrence of surprises should be a warning against ready acceptance of learned opinion. At one time it was thought that the substances synthesized by living organisms were a quite separate set from those that could be synthesized in the laboratory from 'dead' chemicals. Then in 1828 Wohler synthesized urea by heating ammonium cyanate and that marked the death of the vitalistic theory.

The indeterminacy of the degree of truth in our basic models does not prevent the extensive use of composition methods for deciding the relative truth of a novel model; novel in the sense of being a new collection of old building blocks. Even though we cannot assign probabilities to the 'soundness' of each building block, we expect the soundness of the resulting assembled structure to reflect the combined probabilities. In the experimental arrangement shown in Figure 3.5 each vane on the vertical shaft is given a flick to start it spinning. Suppose that each vane occupies a total of 9° of arc. The probability of a small mass dropped from above the vanes passing the first vane without striking it is 39/40. The

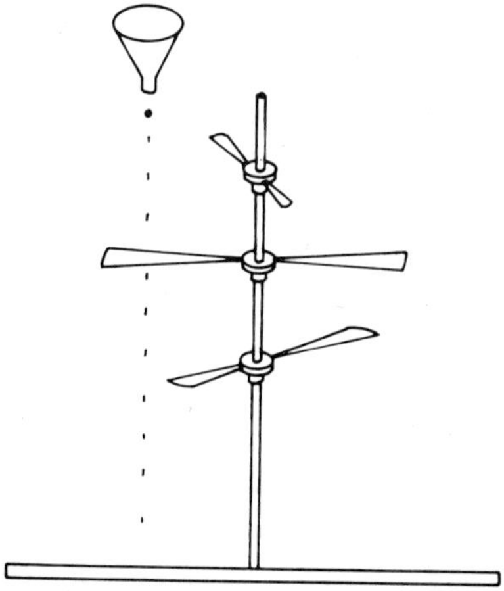

FIGURE 3.5 Model devised to consider the probability of a small mass being dropped from above three spinning vanes, each occupying 9° of arc, striking one of them.

probability of its passing all three vanes in $(39/40)^3$. If no more was known about the vanes except that they were 'very thin' the implication would be that there was a very good chance of the mass dropping through freely. This is the thinking behind large numbers of composition models: assemblies of component parts. As an example, consider the fact that the spectrum of individual frequencies of emission, given out where ions of an element are present in a flame, is observed to show broadening of the lines as seen through a spectrometer. The model for line broadening depends on three components. As an electron in one of the shells surrounding the nucleus of an atom falls between energy states, there is a precise frequency for the radiation emitted. This is the first component. The atoms from which radiation is emitted are not stationary but moving in all directions with velocities corresponding to a calculable distribution. Finally, the frequency of any source of waves as seen by an observer is changed if the source is moving relative to the observer (Doppler effect). Thus with no preferred direction of movement, half the emitting atoms will have a relative velocity towards the spectrometer and will be observed to have higher frequencies, while the other half will be responsible for the lower frequency end of the broadened line.

Provided that all three parts of this explanation are acceptable then the phenomenon is explained. Now the quantum theory, the Maxwell–Boltzmann distribution and the Doppler effect, all enjoy high status and it is unlikely that the explanation will be challenged. Contrast the status of these three building blocks with Bode's law. Bode's law refers to the regularity in the radii of the planetary orbits (Table 3.4).

TABLE 3.4 Bode's law, showing regularity of radii of planets

Planet	Orbit radius	Bode radius	% error
Mercury	3·9	0+4 = 4	+2
Venus	7·2	3+4 = 7	−3
Earth	10 (chosen)	6+4 = 10	0
Mars	15·2	12+4 = 16	+5
Asteroids		24+4 = 28	
Jupiter	52	48+4 = 52	0
Saturn	95·5	96+4 = 100	+4
Uranus	192·2	192+4 = 196	+2

We have no other planetary systems on which to test it because suitable stars which might have encircling planets are too far away for the detailed observation that is needed. No cross checks from other phenomena can be devised because it applies only to planets (and not for example the moons of Jupiter). If it were falsified by the arrival of information about other planetary systems, the text books would not need to be rewritten. Intriguing though it is, its status is that of a curiosity.

There is no necessary relation between predictive accuracy and security. The presence of much noise lowers accuracy accordingly and if the noise level is very high then a model may be dismissed as 'useless' for predictive purposes. Furthermore, the noise level may be so high that the incentive to search for the signal which is buried there is too low for discovery to be made. Some advances in the physical sciences have taken place because measurement technique has improved, the noise contribution of the equipment has been reduced. Security occupies another dimension to that of predictive accuracy. A model may possess high predictive accuracy and yet be local in space or time. Trials with pendulums at one geographical location and altitude enable the time of oscillation of other pendulums found at the same spot to be predicted with great accuracy. A change of height or latitude upsets the prediction. The parameter g, the acceleration due to gravity is not invariant through space. The fact that scientific progress has been characterized by the emergence of parameters with greater generality is encouraging to modellers. The gravitational field strength is local but the parameter G appears to be universal. (The invariance of total mass was found to be local and replaced by the more general invariance of mass + energy using $m = E/c^2$ to carry out the addition.)

There are many forecasting models in day to day use for which the parameters are discovered from a time series and assumed to be invariant until the predictions go seriously wrong. In order to plan the production rate for a given product a forecast of sales volume is needed, and it is demonstrable in a large number of cases that the best forecast, in the short term, is obtained by continuing the sales graph up (or down) the chart. Yet such a model is insecure by comparison with the basic models of physics. It is strictly local, and for a rising sales trend cannot be sound in the longer term. There was a period in which the revenue from air traffic (and the passenger miles) in the U.K. doubled every four years. In a century this rate of increase would multiply passenger miles by more than 30 000 000 times, and in two centuries by about 10^{14}. To demonstrate the absurdity of perpetual exponential growth of anything involving substance or energy or information, we turn to the basic models of physics. (There is nothing contrary to physical law in 10% p.a. monetary inflation continuing for ever, because this does not entail an exponentially increasing requirement for substance, energy or information.)

The paradox of a local model with a good predictive record, and the global model with contrary prognostication, is revealed strikingly in the performance of the gambler's stake-doubling strategy. Faced with a succession of equal probability outcomes, lose or win an amount equal to the stake (for example a roulette wheel without zeros for red or black stakes), the gambler doubles his stake for each of a succession of losers and reverts to his original stake after a winner. Thus each coup ending with a win nets one unit of gain. For example, lose, lose, lose, win, has as an accounting: $-1, -2, -4, +8$ yielding a net $+1$. The average gain is $\frac{1}{2}$ unit per bet. Suppose that the gambler starts with a capital of 1000 units. The typical outcome of this gambling system is a steady unspectacular increase of capital. By the time the gambler has won an additional 23 units he can withstand a run of 9 losers (assuming a no stake-limit house). If he reaches 2047 units he can stand 10 successive losses. Imagine this system offered to lots of punters. The vast majority will be pleased with the first night's trial at the casino. This follows because the probability of 9 successive reds or blacks reaches 0·5 in a run of 1024. It reaches only about 0·05 in a run of 100. The global model for this gambling system, and for all gambling systems facing an unbiased wheel with finite capital, yields a prognosis of ultimate ruin. The probability of doubling the capital is exactly equal to losing it all for the system above and for all systems (including those not yet devised).

The language in the latter part of the preceding paragraph is not qualified in the way statements about physical models should always be qualified. The statements follow directly from the specification of the context. They are mathematically, that is tautologically, true. In this sense the version of the second law of thermodynamics which states that the entropy of a closed system can only increase, is a tautology. This is not to denigrate the remarkable insight which led to the formulation of the second law. It may require great skill to uncover the full implications of a small set of propositions; or to assemble a model with interesting properties from simple and well-known components.

It is tempting to suggest that the second law of thermodynamics is true because it depends on the mathematics of probability. However, its truth has meaning only in relation to real systems. There is no way of being sure that the universe, or even some selected part of it, is a closed system. This inquiry must end somewhat unsatisfactorily. The models of science are a set of working hypotheses. The probability of any one of them being true cannot be calculated. But they can be ranked according to how local or general each one is in its application, and the quality of a local or general model can be measured under defined conditions in terms of its predictive accuracy.

4. Control

Purposes of modellers

Models are constructed for at least three different reasons. The modeller can simply take pleasure in the activity of drawing order and elegance out of apparent chaos. The satisfaction here is akin to that of assembling the pieces of a jigsaw puzzle. It can be argued that solving Nature's puzzles is a 'higher' activity than doing *The Times* crossword, because the solutions to natural riddles are often strange enough to need high intellect in the solver. As a G. K. Chesterton character remarks: 'Fact is of necessity stranger than fiction because fiction is a creation of the human mind and therefore congenial to it'.

Some models, however, are initiated by a desire for change. The purpose is prescriptive rather than descriptive. Under certain conditions of air flow over the wings, aeroplanes stall. An aeroplane which stalls at low altitude is very likely to crash. It was imperative, at an early stage of aeroplane development, to find good models of air flow in the neighbourhood of aerofoils and thus to understand the change from laminar to turbulent movement of the air. The advocate of political change offers a model of the socio–economic system with the aim of persuading his audience that the prescription he offers will result in a change of state of the system that is desirable. Finally, there is a third type of modelling which is neither disinterested nor aligned with specific goals. Most occupations have aims which are clear and unambiguous: to carry out the task in hand more efficiently, or cheaper, or quicker or more enjoyably; whether the task is raising pigs or teaching arithmetic, or catching parking offenders. Now consider the problems facing the manager of a restaurant. He knows that the staff engage in various practices which defraud the customers and the owners. They do this sometimes alone and sometimes in collusion with each other. He improves the accounting procedures and divides the responsibilities so that it becomes difficult for the staff to thieve without being detected. Then he finds that the waiters are leaving and taking jobs elsewhere. He discovers that the market rate for catering labour is set against a background of the 'normal' rate of perks. If the perks are stopped then the job appears underpaid. This is the time for an agonizing reappraisal. There are some situations for which the appropriate goal only emerges through a fuller understanding of the constituent parts of the system. The modelling activity appropriate to these situations is purposive—it is called into being by some feature that was called a 'problem'—but it is not a mere rationalization of a course of action already decided (like the models presented in most political tracts), rather it is an exploration with the goal itself being open until the model is sufficiently comprehensive. This could be termed the proper purpose of Operational Research; the seeking of good goals through better understanding, together with the effective means of implementing them. Whether the new regime can be implemented takes us into the larger realm of control. Since this concept of control is central to any discussion of systems we must consider it in some detail.

The meaning of control

For many of the sentences in which the word 'control' is used its meaning is clear. If we speak of a driver being in control of a car, the inference is that through his movements of the steering wheel, accelerator, etc. he maintains the state of the car within specified ranges of certain variables, for example, position on the road, speed, engine speed/road wheel speed ratio, illumination area ahead. This is the sense in which a thermostat is said to control the temperature of a room, by maintaining the temperature within a specified band. A slight ambiguity enters if the reference is to a car being 'out of control'. This can mean either that the driver is asleep/intoxicated/absent or that the car is not controllable because for example, it is skidding on a patch of ice. In practice an operational definition is used. The observation of a state of the car for which one or more variables lies outside the specified range is taken as an indication that the car is 'out of control': the distinction between automotive and neural mechanism is ignored.

When the system is enlarged the clarity recedes. Take the following statements:

1 The fire has been brought under control;
2 The army is now in control;
3 He saw red and lost all control;
4 Man is now a slave to the machine; he is controlled by it.

For purposes of precision, the discipline of control engineering prescribes an exact meaning for controllable. Given a set of state variables defining a system and a set of control variables which affect the state, then the system is controllable if all changes between one state and another can be achieved with the given controls. It is conceivable to apply this definition to the first statement. Using the variables (quantity of water, etc.) the fireman can decide the progress of the fire, including its extinction. The second statement is more obscure, because in the case of a political coup, there are widening circles of people aware at later and later times that the coup has occurred. If communications are poor, the lapse of time can be days or weeks. The people who learn of the coup late are not affected at all during the intervening time. There is no way in which the presence of control could be demonstrated for these people. It is true absolutely that they are not controlled. The army may make no changes in administration, in which case there is no objective evidence of control, but there may be belief in the potential for control. This is a far cry from the control engineer's definition. All persons in authority have good reason not to attempt to discover the ultimate limits of their power because if the experiment flops then belief in their authority is undermined. The second statement should be revised to: 'The army has some influence whose extent is unknown'.

In the third statement, the unspoken assumption is that individuals are like machines and can cease to be controllable in the way that machines go 'out of control'. This in turn implies a controller of the human machine; an entity operating at a higher level than the emotional. In practice it may merely refer to any violent behaviour

The fourth statement is the most interesting departure from the control engineer's usage. It raises the question of who or what is the control in situations containing joint inputs which are both requisite to achieving changes. In all group activities where the participants play roles each of which contributes to the outcome, all that can be stated is that no-one has full control because he depends on the co-operation of the others and that everyone has much more limited control of veto. We distinguish various levels of influence among the group members. The 'leader' may be accorded the option of inflicting penalties on unco-operative members. He may be recognized to have superior strategic skill and the members agree to his recipes. He may be a persuasive advocate and able to convince the others of a given course of action. The difficulty of reaching an objective judgement on levels of influence comes about because leadership is itself a role. The leader must conform to the picture that the group members hold of acceptable leaders. His strategies must lie within the set of acceptable strategies. Examples abound of individuals who were accepted as leaders only in a particular context. One is reminded of the queen bee in the hive. Provided that she continues to lay eggs and to exude that strange substance which, passed to all hive members reassures them of her vigour, then all is well. The worker bees will forage, the guard bees will fend off intruders and the nurse bees will tend the incubating eggs. If for any reason the queen's activities are impaired, the hive generates new queens; only one of which will survive to perform the regal functions.

It may be that a visitor to this planet could describe all human behaviour in the way that we can describe the beehive, but our own awareness of the loose analogy between the hive and human groupings brings in the dimension of conscious choice. The structures of human societies have, in various times and places, spanned a wide range of hierarchy acceptance. At one extreme, collective decision-making in small tribes achieving near complete consensus and at the other, rigid authoritarian pyramids of power. For any persistence of structure there must be a shared view of the way that the members inter-relate: a shared model of the society and a shared world picture. From time to time there can come about changes in this shared world picture, and these changes supply an unending source of material on which historians can weave their own patterns of apparent cause and effect. Without entering the never-ending debate on the sources of historical change, it will be agreed that the wide-spread acceptance of a new world picture can wreak profound changes. If it can be further agreed that there is a key role for the authors of new world pictures, then a way of comprehending the concepts of influence may be reached. The next step of describing how history was changed by Marx, or by Keynes is easy but simplistic. The author may strike no chord in his audience; he can be 'before his time'. Like the leader who needs the approval of the led, so the sower of new seeds needs fertile ground. With this reservation about the opportunities for seeding minds with new world pictures, we can note that a special class of seeds enjoy a high germination rate. New pictures of the physical world from researchers in the physical sciences are readily accepted. Provided that demonstration is possible through the replication of supporting experiments, and that consistency between the new picture and other older established pictures is apparent, then the scientific community incorporates the new picture so that it augments and embellishes the old.

Adaptation

Refining its representations of reality is not confined to the scientific community. In a stark form it is a pre-requisite for the survival of any organism. The learning process by which living creatures develop responses to their environment that enhance survival correspond to the improving of the match between inner form (within the creature) and outer form (the creature's environment). This theme has been developed in detail by W. Ross Ashby in *Introduction to Cybernetics*[6] and *Design for a Brain.*[7] Ashby has been able to show that the structure of organisms or mechanisms which show properties of adaptation, necessarily entails two separate levels. The first level is that of immediate direct response to environmental change. The simplest example in mechanistic terms is that of the thermostat responding to a temperature change. However, the response itself is a function of the parameters defining how the thermostat has been designed and set for regulation. Clearly the alteration of the temperature setting on the thermostat is of a different order from that of alteration in the environmental temperature to which the thermostat is exposed. This setting operation belongs to the second level.

As Cantley[8] has observed, the two levels, often displayed as an inner and outer loop conjoined, have been invented several times by different authors to describe the adaptive behaviour of organizations. 'Invented' conveys overtones of superimposing a pattern where nonè exists, and this fails to do justice to the power of Ashby's exposition. He has shown on strictly logical grounds that the identification of adaptive behaviour is, in a strict sense, also the identification of an adaptive mechanism, a proof of the existence of the second loop. At the lowest levels of single celled creatures we can observe the novel patterns of behaviour called forth by exogenous inputs to their surroundings. As we climb through the levels of greater complexity in the organisms, the types of adaptive behaviour multiply, but the principle is unchanged. It is not a particularly big step from the learning and adapting nature of single creatures to the same survival-enhancing properties of creatures in groups or colonies. Though the details of the mechanics have yet to be discovered in many cases, the macro behaviour can be examined and replicated in the laboratory and the field. The external observer, ignoring the potential for shared experience with the actors, can discern this same outer loop adaptive facility in human organizations. But additionally in human group adaptation, there appears the ability to communicate the adaptive structure formally: to rehearse the operation of a novel structure as a mental model. The new construct may reflect a poor representation of reality, in which case its implementation will fail or the results persist only briefly. This is the same selective process that is evident at more primitive stages of the evolutionary ladder.

The picture of adaptation as a group property, drawn most clearly from biology, is a most fruitful means of approaching the related ideas of modelling, control and historical change. The notions of status and hierarchy are so strong in our culture that there is a constant seeking for the 'key' elements of power in any given situation. An alternative viewpoint is that of regarding the total behavioural pattern of a group as the manifestation of an adaptive structure, in which the several elements of adaptive mechanism can be identified with the several roles of the actors. No single actor can perform the play unaided. It is not a sensible question to ask which part of Ashby's homeostat is more important than any other, because its behaviour is a consequence of the totality of its parts and the connections between them.

The 'normal' behaviour of the group can be described as a stream of transactions following formal or informal procedures, with familiar roles being played. The individual actors who comprise the cast may change and bring their own styles to the parts, but the plot is the same. From time to time the play loses its appeal and a new one is needed, or at any rate a variation on the old theme. The playwright may or may not get his new creation accepted. Even if it is accepted the performance may flop. Some of the offerings will be dubbed too *avant-garde*. There will be critics to complain that the new play is only the old in modern guise. . . .

Analogies

The scientist who turns from the laboratory and the physical sciences to become involved with systems of people is like the manipulator of puppets trying his hand at plays in a live theatre, and discovering that the change is fraught with the hazards of unwilling backers, uncomprehending producers, truculent players and a capricious public.

The mixing of language appropriate to the theatre with that to describe adaptive behaviour is deliberate. Ashby's first control level: the retention of variable values within restricted ranges (repeat performances of the old play) is easily understood. For his second level of control we may admire the apparent purposefulness of his randomly juggling homeostat but we believe in the real purpose of the new regime (a new play and not a random mixture of parts of old plays). In the larger context of the theatre company, the playwright, the producer and the backers are also role players along with the cast. All the roles are necessary, for a failure of any one of them results in overall failure. The performance takes place in a setting, that is, subject to constraints of its environment, and it must convey *something* or else it fails; it is non-viable. The term 'control' for the second level suggests an agent, a controller: but the reality of the control agent must be an open question. The biologists come near to describing all of the evolutionary advances of species in terms of self-selection through superior adaptation to the environment. Some plays are successful and some societal structures endure. There are constraints on possible organisms and on possible societies. Within the boundaries set by possibility you can describe the rich profusion of life forms as the blind play of chance or you can invest it with significance. Societal structures are bounded but diverse in form. Control in the form of the boundaries is present but within the bounded region there can be variety: through chance or artistic creation, whichever you prefer. The individuals in groups —in institutions, organizations and societies, play through the various roles—some concerned with the basic machinery, some with the adaptive needs, some hastening demise of the current structure so that something supposedly more viable can succeed it, and some studying the boundaries and viability *per se*.

The analysis of second level control through bounded envelopes of possibility can be carried out by the methods used in the sciences. In contrast, models of 'history' (with the concomitant forecasting of the contents of future history books) appear as likely as evolution models that would reveal say the peacock and the liver-fluke as logically necessary to the scheme of things. Such a feat cannot be ruled impossible but it must be judged unlikely.

Control and manipulation

The extensive use of analogy in this discussion of control may assist in distinguishing the levels, and in appreciating the different sorts of models which are appropriate to each of the levels. The distinction is important, because models at the first level are familar and frequently useful in a direct fashion for prediction or manipulation, whereas at the second level the significance of the models is not in point predictions or in the provision of specific advice to manipulators. This does not mean that second level models possess only scientific elegance. The general circulation models of the atmosphere are destined to tell us much about climatic possibilities (though little about day to day weather forecasting). In a more direct and personal way the two levels appear strikingly as one contemplates the emergence of a new process, industry or preference. In Figure 4.1 are shown plots of substitution phenomena taking place. They all conform closely to a logistic relationship: the logarithm of $f/(1-f)$ where f is fraction of the market, is a linear function of time. It appears that after penetration has taken place to the extent of about 10% the rate of further substitution is predictable with quite small error. The examples shown are due to Fisher and Pry.[9] The takeover of coal from wood, and oil from coal has been shown by Marchetti[10] to conform to the same logistic relationship.

Imagine the situation of a director in a company whose product is being supplanted by a new one. He can advocate fighting back—or he can preside over the demise of the old firm—or he can propose that the board, because it cannot beat them, should join them and transfer to making the new product. The directors do not agree and a minor drama is played out. Whole novels have been written on lesser themes. To the observer plotting the share of the market as the sum total of many firms, each with its own decisions, all the local politics is quite insignificant because he can see the broad sweep of change taking place.

If we take just one substitution, that of wood by coal, we can see that there is nothing inevitable about the choice. The industrial revolution did not have to be fuelled. The fact that charcoal for

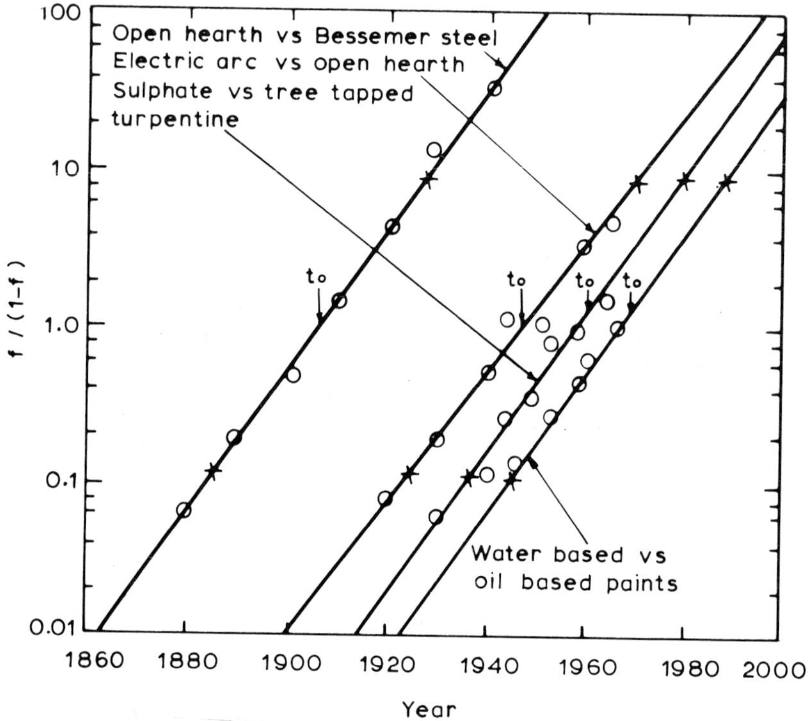

FIGURE 4.1 Plots of substitution phenomena at four different periods of time. All show a remarkable consistency of further penetration of the market after the first 10% entry had been established (Fisher and Pry).

smelting was causing the thinning of forests could have prompted technologies not dependent on iron. One feature only is sure, that the situation demanded an adaptation: a new style that could persist without a growing appetite for wood. Substitutions in the market are like the shifts to be seen in species populations. White moths give way to dark moths in the woods that border smoky industrial areas, because the white variety is more vulnerable to predators. There is a temptation to regard these sorts of shifts as predictable: but this is only with the benefit of hindsight. As for the coal example, there are many potential new viable styles, as demonstrated by the different ways in which societies in different parts of the world have met their crises of adaptation in different ways. Yet the new forms are not wholly unknowable because only a limited set of all conceivable structures (or creatures) can replace the old. It is in this sense of a bounded class that the second level is amenable to study.

Adaptive models

All of the example models considered in earlier chapters are structures envisaged in particular situations: they were sometimes specified but occasionally tacitly assumed. The trajectories of bodies in the solar system are closely predictable, provided that some large object comparable with, say, the mass of Jupiter does not arrive from outside and exert a new gravitational influence. The specification for Nicholson's blowfly population to execute its clockwork-like cycles was quite precise. Meadows' hog-cycle operates within a particular set of economic rules; these derive from the given societal setting. These are all, what we may call, models of the first type: to correspond with the first level of control. The ideas of determinable parameters, background noise and predictability apply to these models because they have stationary structures. Of course, in laboratory work the object is to ensure as far as possible, a stationary, unchanging structure throughout the course of the experiment. Outside the laboratory, the evidence of a long sequence of observations conforming with a plausible structure containing few parameters is used to infer a stationary structure. In the physical sciences, the use of stationary structures is pervasive. For example, even if our solar system were to be invaded by an alien large mass, it would be possible by applying the general gravitational law to establish, after a period for observations, the new trajectories and new predictions for planetary positions. There would have come about a variant of essentially the same basic structure. In Figure 4.2 are drawn a series of structures in topological form; only the linkages between the nodes are used for distinguishing between separate structures. The variations of linkage length have been made to suggest changes to the variables that define the state of each system. The meanings of the

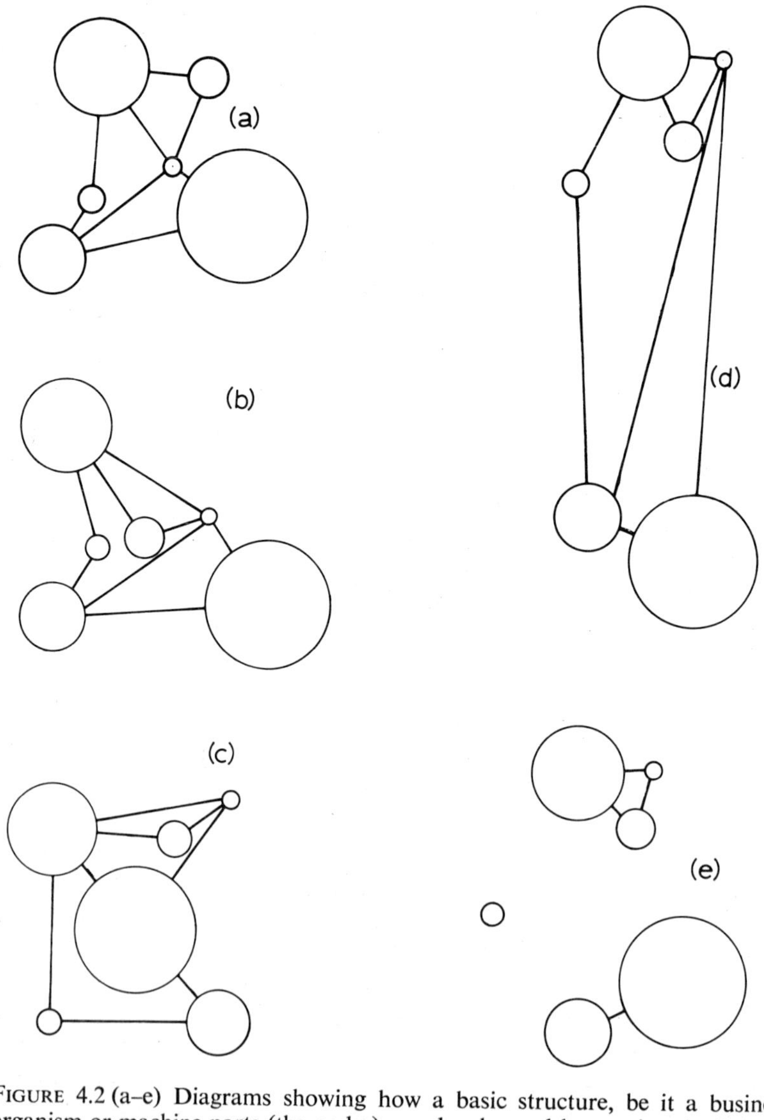

FIGURE 4.2 (a–e) Diagrams showing how a basic structure, be it a business, organism or machine parts (the nodes), can be changed by varying the linkages between the nodes. The structure of 6 nodes remains the same but the systems can be vastly different.

nodes and the arcs can be chosen according to the system and to the description of it which is conventionally employed.

For a firm, the diagram can consist of nodes representing the individual functions of production, sales, accounting, dispatch, wages, stores, personnel, etc. and the arcs, the information, material and money flows travelling between the functions. For a living organism like a moth, the diagram identifying each separate organ and the flow lines of influence between each would be elaborate, but in principle it is possible to construct it.

The idea intended to be conveyed by (a)–(d) of the diagrams in Figure 4.2 is that of changing states of the system while the structure remains the same. The causal linkages in Meadows' model persist even though the U.S. administration changes and the economic climate surges to and fro. The functions in the firm continue during the normal vicissitudes of business life. The moth's metabolic processes and its sensory–motor feedback mechanisms carry on working throughout its lifetime. Returning to Figure 4.2(d), there are symptoms of extreme values for one or more variables and in 4.2(e) the structure has changed, or put more colourfully, has disintegrated. A firm going bankrupt and its employees departing to other jobs represents a clear change of structure (for the firm regarded as a single entity). A moth eaten by a predator persists as substance but the integrity of its structure is lost. While the original structure of a system persists we may tackle the task of understanding it and even predicting its state through models of *Type 1*. When it adapts by changing its structure or the elements of it become separated and rejoined to form a new structure, then a different sort of model is needed: a *Type 2* model.

One way of thinking about drastic structural changes, like bankruptcy or death, in populations of firms or moths is to disregard the individual members of the populations and consider only the statistical properties of the ensembles. This is typified by the approach described earlier, which has been used by Fisher and Pry in their studies of substitutions occurring in products or processes. Similarly, in the blowflies study, individual flies as autonomous functioning entities are ignored and only their number at each time-interval is recorded. Thus, this way of modelling amounts to reducing the study to a *Type 1* model; at the cost of losing any insight into structural change as such. It is clearly not adequate for either more subtle structural changes than disintegration nor for cases that do not fit into the context of an ensemble. Organizations can change their structure without disintegration and organisms can adapt to threats so that they avoid prompt extinction.

Ashby has proposed that adaptive activity in organisms can be explained by a random searching among the control settings of the outer loop, coupled with a reinforcement of remembered successful ploys. It is of interest to note that these two ideas have been employed deliberately by management science exponents. Faced by a new situation it is often recommended that the first step should be a brainstorming session at which all the possibilities that any of the participants can think of (including the most unlikely) are aired. The second idea of calling on experience is the stock-in-trade of management consultants, who in consequence of observing many organizations are able to match the symptoms of their current case against similar cases (and cures) previously encountered.

Brainstorming and the distillation of experience are the participatory part of *Type 2* models. There is an active role in outer loop control: devising new procedures, merging or splitting functions, changing the information flows; in the earlier analogy, writing new plays. There is also the role of observer and modeller in providing a conceptual form for structural change. This form is distinct from the *Type 1* model with stationary structure and intrinsic predictability. Even a few elements comprising a structure yield an alarmingly large number of theoretically possible linkages. Five elements can be linked in over 15 000 distinct combinations. However, there are powerful constraints in real situations which drastically reduce the number of combinations that need to be considered. Structures with the qualities of resilience and persistence must have stabilizing, that is negative, feedback. It is no accident that Meadows' hog-cycle model is composed of two negative feedback loops. Some positive feedback loops in a persistent structure can occur but the over-riding influence must be self stabilising. For example, a rise in demand following a rise in price is a positive feedback; and does indeed occur in some isolated markets over small ranges, but it could not persist generally for any considerable length of time. Secondly, it can be shown that a high degree of connectedness is associated with a high probability of instability. Connectedness here refers to the ratio of number of links to number of nodes. Real functioning systems do not have link–node ratios greater than about 2.

The two constraints noted above are internal. An additional set of constraints arises in the relation between the system and its environment: the stability requirements operate here also. The organism or organization must react with its environment in ways that restrict the ranges of the variables. At an elementary level this requires, for example, that warm blooded animals possess mechanisms causing greater than normal heat loss when a rise of temperature of the environment occurs.

When all the constraints have been applied to reduce the variety of hypothetical structures to those which are potentially successful adaptations, this narrowing down does not, in general, supply the predictive quality of a *Type 1* model. Indeed by its very nature, this emergence of a changed structure will occasionally manifest as the creation of a unique entity, without forerunners which are remotely similar. It may be objected that new organisms only appear as the result of many small changes in the course of evolution, or that the emergence of new structures for institutions and organizations occurs almost imperceptibly over time. Although this 'gradualist' view appears plausible, it overlooks the problem of the necessity for stability in the structure which preceded the change. Merely to assure survival, the preceding structure must react to the host of internal and external disturbances without changing its structure. Then there comes a time when the elastic is stretched too far (not inapt because springs and elastic in mechanical structures perform as elements in negative feedback loops) and a new form appears; or as an alternative description, the old form 'dies' and a new form enters.

Control and type two models

Controlling the *Type 1* structure is a matter of routines and procedures. The actors are themselves elements in the stabilizing feedback loops. If someone describes himself as a mere cog in the machine, this is not just picturesque; his role may be as clearly specified and as closely defined as that of a component in a machine. The meaning of control when adaptation is needed is so different from the *Type 1* case that one needs new words. A range of situations must be considered. At one extreme, the constraints on possible adaptations are so severe that the outcome is decided by the context; the environment of the organism or organization. Light colour moths have no ready means of camouflage; their extinction has an awful inevitability about it. The 'controllers' of companies producing a product with a declining demand can feel themselves prisoners of the situation. At the other extreme we can note that a single individual with a vision of an alternative future, combined with the power to imbue those around with enthusiasm to make the vision a reality, can generate a step change. Much more common are the intermediate cases where the environmental influence is strong but not decisive, there are visionaries but they are disregarded, and to quantify the relations between change factors and outcomes is beyond our power.

At the first extreme of strong environmental influence the modeller can use *Type 1* models and predict with useful accuracy. In the middle range, he can narrow the spread of outcome, and this too is a useful contribution to the actors in the scene. He can also be a catalyst because the display of potential outcomes itself influences the actors. Thus the *Type 2* situation is truly of a different order. The construction of *Type 2* models (unless this be done very privately) itself contributes to the outcome. There is a parallel here, between classical physics and post quantum theory physics. Before quantum theory we could regard the observer as quite separate from the observed (*Type 1*). Now we know that at the electron level, the observations themselves influence the observed quite unavoidably (*Type 2*).

5. World models

In recent years there have been attempts to construct models embracing human activity over the whole planet. These enterprises have aroused wide interest and heated debate. It is pertinent to ask the same questions about these models as one would ask about models of any other system. For what purposes are they constructed? What parts of the system are represented? Are the models intended to yield point predictions in time, or distributions with measurable characteristics? Are there means of checking them by reference to experiment or observation? In the language of the last chapter, the question arises of using *Type 1* or *Type 2* models for thinking about the future of the planet and of the measurable characteristics of human activity which might be inferred. There are attractions associated with *Type 1* models. If it were possible to account for observations of past values for the variables of the world system by means of simple relations (essentially independent of the observer), this would be a notable scientific success. We can already explain and predict tidal motion. Recently there have become available data on climate stretching back over hundreds of thousands of years, from which it seems likely that the causative mechanisms for ice ages can be identified with cyclic variations of earth orbit and axis inclination. These cyclic variations follow from well understood gravitational effects, so that prediction of the next ice age (not necessarily with high precision) could follow.

It has been suggested that simple functional descriptions can be devised for some of the variables that apply to the global human population. Figure 5.1 is a plot on a logarithmic scale of world population. The curvature of the line indicates that the rate of

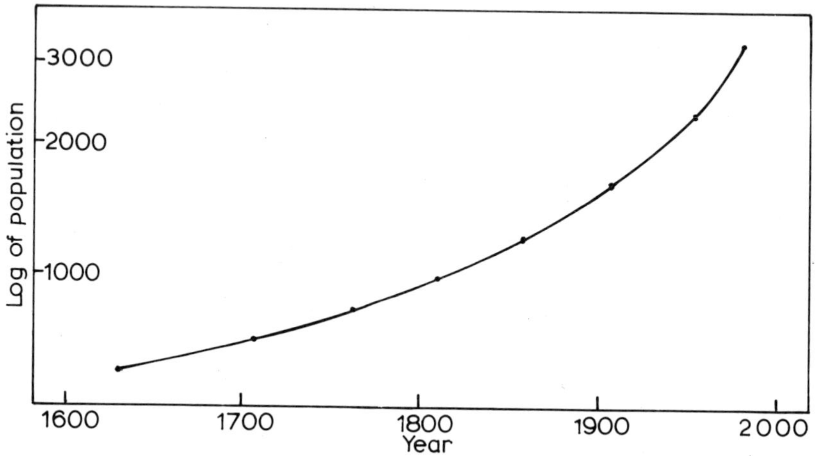

FIGURE 5.1 Plot, on a logarithmic scale, of world population indicating an increasing growth rate.

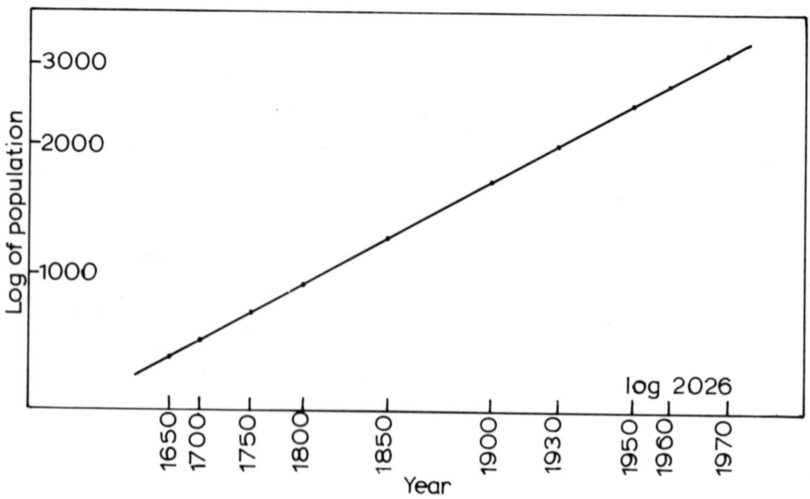

FIGURE 5.2 Plot, on a double logarithmic scale, of world population, with a time origin of $t_o = 2026$, showing a linear growth rate (Meyer and Vallee).

TABLE 5.1 World population growth rate from 1650–1972

Year	Population (millions)	Yearly rate of growth (%)
1972	3782	
		2·03
1971	3706	
		2·02
1970	3632	
		1·97
1960	2982	
		1·82
1950	2486	
		0·91
1930	2070	
		0·86
1900	1600	
		0·62
1850	1070	
		0·52
1800	900	
		0·44
1750	720	
		0·31
1700	620	
		0·27
1650	540	

growth has itself been increasing. Table 5.1 of growth rates during the past three centuries shows this clearly. Meyer and Vallee[11] pointed out that using a double logarithmic scale with a time origin of $t_o = 2026$, the plot is linear (Figure 5.2). The equation of this line is:

$$\log P = 11{\cdot}310 - \log(2026 - t) \tag{5.1}$$

The expression for world population as a function of time is:

$$P = \frac{2 \times 10^{11}}{2026 - t} \tag{5.2}$$

A more striking way of presenting the same data is to plot area per person available as time passes. Figure 5.3 shows that

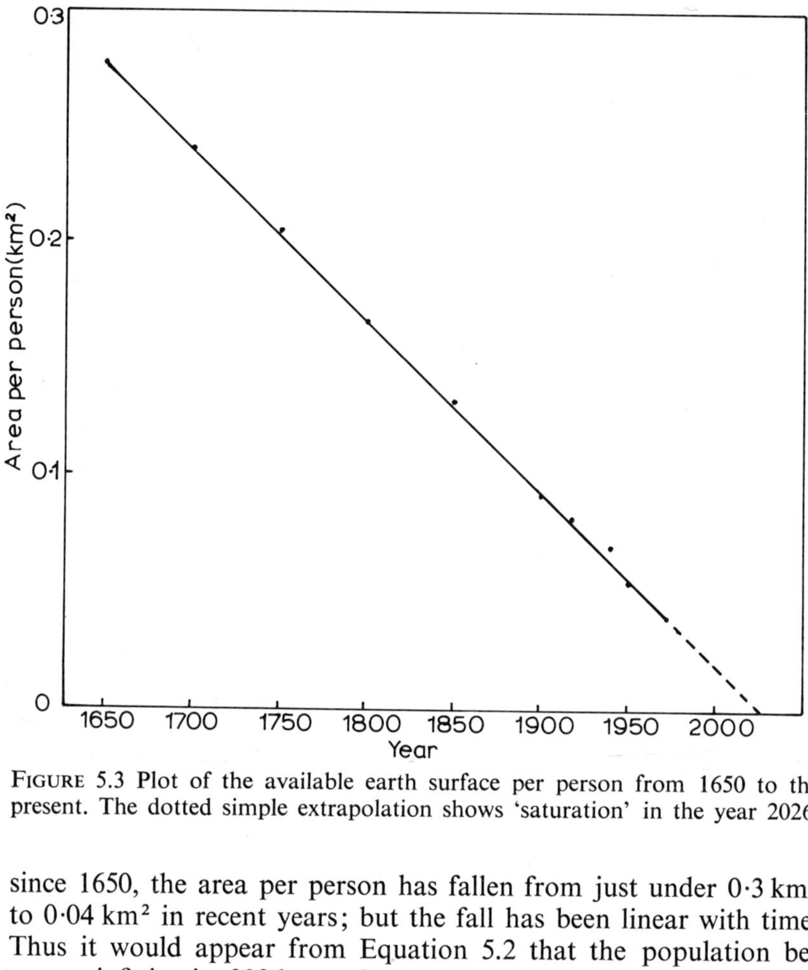

FIGURE 5.3 Plot of the available earth surface per person from 1650 to the present. The dotted simple extrapolation shows 'saturation' in the year 2026.

since 1650, the area per person has fallen from just under 0·3 km^2 to 0·04 km^2 in recent years; but the fall has been linear with time. Thus it would appear from Equation 5.2 that the population becomes infinite in 2026, or alternatively from Figure 5.3 that the area per person becomes zero in the same year! Recalling that consistency, parsimony of parameters and predictive power are the hallmarks of successful models, there might appear to be a paradox here. A model fulfilling the required criteria generates a manifestly absurd prognostication. The resolution of the paradox is not obscure; first, the relationship is a mathematical description and not a model, secondly, the history of human population increase during the last few thousand years is a single isolated phenomenon without comparison from other epochs or other planets. The nearest comparison is from animal populations which, though prone in favourable locales to explosive growth, do not sustain it for long. The remarks about Bode's 'law' in Chapter 3 apply with equal force to this population growth curve fitting exercise.

Extrapolations, unless justified by a supporting explanatory model whose elements are drawn from the interleaved web of well established scientific laws, occupy a lowly place in the order of prediction methodology. It has been remarked previously that extrapolating sales graphs into the future is often the best known way of estimating demand. This does not contradict the preceding statement above concerning the weakness of extrapolation. If the costs which will follow from mistaking the sales trend are not excessive, then sales forecasting by extrapolative methods for the next few months or years is a sensible practical strategy, because a business cannot be conducted on the basis of requiring precision in all the information used to plan operations.

Of a quite different order from trend watching is the assembly of structures whose rationale is open to discriminatory testing. If the interest is in human population dynamics then it is appropriate to investigate it through the study of changes which have occurred to human groups in various times and at various places; also through the study of similar animals such as the other primates. Malthus offered a model of human population growth which can be examined by the criteria which apply to any model. The Malthus postulate was that human fecundity is persistent (and therefore continuing population growth inevitable) coupled with a finite limit on food production. Specifically he argued that a geometric population increase is accompanied by only an arithmetic increase in food output. The restraints on population growth are thus famine, but also war and disease.

Supporting evidence would be provided for the Malthus thesis if it could be shown that wherever wars or disease are not efficient limiters, then a population rises to that level set by food output and exists thereafter at the edge of starvation (possibly overshooting and recovering like Nicholson's blowflies). Some corroboration for the thesis would be obtained if it were known that tribes of baboons live at the edge of starvation (because baboons, of all the primates, come closest in social organization and life style to our recent ancestors). If there were no signs of feedback effects from population density to fertility—so that birth rate tended to fall when the resource base appeared limited, and rose in times of abundance—this again would lend support to Malthus.

It has to be said that on the three points noted above, the data which are obtainable do not provide unequivocal support for Malthus. Individual, relatively isolated groups living in South America, Australia, Africa, Brunei and New Guinea appear not to press constantly against food limits. They practice a variety of means of inhibiting runaway population growth. Many animal species, including baboons, appear to react to the signs of over population by reducing the number of their surviving offspring. The Malthus model is simplistic, or in the terms used in Chapter 3, the model has local rather than universal application. Some species certainly multiply up to the food limit, or to the limit set by their natural predators; but these species are, in biological terms, distinct from the primates, for example, they have rates of reproduction orders of magnitude greater, and minimal care for their young.

Simple extrapolations and unsubstantiated hypotheses are poor bases from which to think objectively about the world's future, and the question arises of what equipment lies to hand if one wished to do better. The first point to recognize is that any model which might be constructed will be dependent on the soundness of the elements in the structure, rather than on acquaintance with many similar cases drawn from examples across space or time. We do not have the records for any other planets where a species like our own has passed through a phase of population and technological growth. Nor do we have past history containing instances of four billion people on this planet attempting to live rewarding lives. The problem can be compared with the example offered in Chapter 3 for modelling the broadening of spectral lines. The three elements—the quantized energy step, the velocity distribution of the atoms and the Doppler effect are secure in the sense used when the nature of scientific laws was discussed. The resulting effect of line broadening is itself as secure as the constituent parts, which in combination, yield the result. Likewise, if individually secure mechanisms can be assembled to model the world system then there can be confidence in the model findings of a strength corresponding to the security of the constituents.

Because the world contains a great variety of chemical compounds, species, climates, societal styles, etc, it should not be assumed that the models used to represent it must in their turn be highly elaborate. The order of complexity needed in the model depends on the variables about which interest is centred. For example, the blowfly is by the standards of man-made laboratory apparatus, an extremely complex entity, yet a model of the fluctuations in a sample population of blowflies can be both remarkably simple in structure and creditably accurate in predictive power. If only a few variables of the world system are of immediate concern then there is no *a priori* reason to believe that an elaborate model is needed. A relatively simple world model, with only a handful of variables and stationary structure (*Type 1* model), was assembled by Professor J. Forrester and published in 1972 under the title *World Dynamics.*[12]

Forrester's model

Human beings have created art forms, reared up great philosophical structures, developed scientific method and fought wars. None of this appears in Forrester's World 2 (as he called it). The inhabitants of World 2 have the following attributes:

1 They reproduce themselves;
2 They die;
3 They eat food;
4 They make capital investment (buildings, machines, etc);
5 They consume material goods.

It can be inferred that human labour is necessary to operate and maintain the machines, though the labour force is not itself quantified in the model. This sort of simplification is characteristic of World 2; for example, the natural resources (all the materials obtainable from the earth's crust) are conceptually aggregated together as an undifferentiated mass which is necessary for production. Only four kinds of physical output are identified, food, material goods, capital equipment and pollutants. These outputs are all treated as homogeneous; even the pollutants are the notional sum of all polluting substances.

Before any criticisms are raised of the simplifications that Forrester used, we should consider the objectives of this model. Growth of population, of resource exploitation, of food requirements and of pollutant generation, have all been occurring at relatively rapid rates (doubling times of only a few decades) within a system boundary (the earth's surface) which is clearly finite. Elementary calculations show that some aspects of growth, like energy dissipation and agricultural output, cannot continue at the same rate for much longer and it is of interest to understand the ways in which the system could adapt. The blowflies adapt to a finite supply of liver by a sudden drastic reduction of population; but this is not the sort of adaption that we would welcome for human societies even if, like the blowflies, the population level recovers again later.

For the purpose of answering the single question of whether there is likely to be overshoot and collapse or, alternatively, a growth slackening and stability, it is possible that a quite simple model with few variables is required. Even if the findings reveal a need for more complexity before a good representation is achieved, there are good arguments for starting with simple structures and only making them more elaborate if the need is demonstrated. The history of science is replete with examples of seemingly complicated behaviour deriving from the interaction of relatively few intrinsically simple elements. Large models with many interactions are unwieldy; it is difficult to provide all the quantitative relations needed for their operation; it becomes an onerous task to ensure that all their possible modes of behaviour have been exhaustively explored.

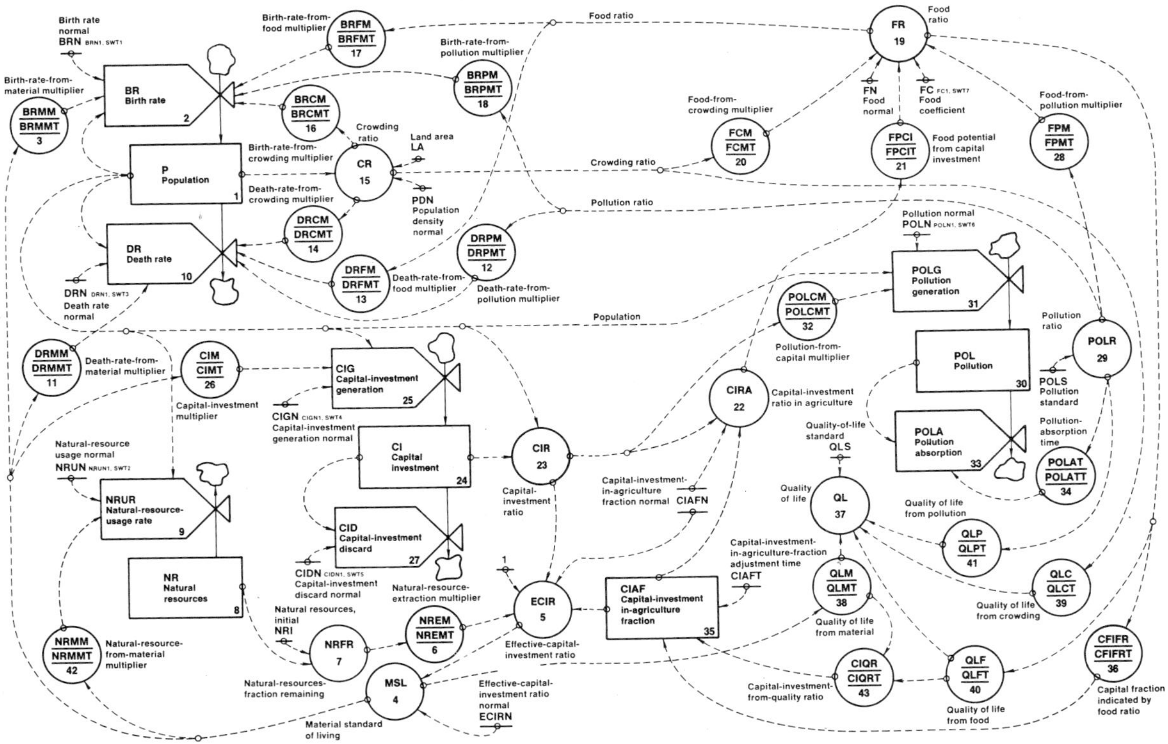

Complete diagram of the world model interrelating the five level variables — population, natural resources, capital investment, capital-investment-in-agriculture fraction, and pollution.

FIGURE 5.4 Complete diagram of the world model, interrelating the five level variables; population, natural resources, capital investment, capital-investment-in-agriculture fraction, and pollution (Forrester's *World Dynamics*).

World 2 Structure

The assembly of the World 2 structure and the justification for each of the relationships used in it have been very clearly and comprehensively described in *World Dynamics*. Rather than attempt a summary of the book, this section will describe the mechanics of the model, and the next, the causal relationships assumed to operate in the real world, and their manner of representation in the model. Figure 5.4 shows the causal link diagram. The sets of rectangle shapes plus one or two valve like shapes which appear four times associated with population, natural resources, capital investment and pollution correspond to accounting operations. For example, in one discrete time interval (chosen to be 0·2 year) the birth rate in births per year × 0·2 is added to the variable 'population' and the deaths per year × 0·2 are subtracted from 'population' so that this variable at the end of the time interval has been changed by the corresponding net increase or decrease. This same procedure is simultaneously applied to the other accounting relations. The sum total of natural resources is reduced by the amount necessary for the material standard of living and the capital investment. The capital stock is increased by investment and a subtraction made for depreciation. The pollution level is increased by a quantum of pollutant derived from the production process and decreased by absorption through environmental biochemical processes. A fraction of capital investment is diverted to the capital stock in agriculture.

In addition to these accounting operations the 'rates' like birth rate, death rate, pollution absorption rate, etc. are changed in their values before they are used in the following time interval. The changes in the rates are made to depend directly or indirectly on the variables subject to the accounting operations (the so-called 'state variables'), population, natural resources remaining, pollution level and capital stock. Thus the births per year and deaths per year are made to depend on the population itself. The ease of extraction of the next quantum of natural resources is made to depend on the quantity remaining. Hence as the time intervals succeed one another, the state variables rise or fall and the actual rates determining these increases or decreases themselves change, thereby yielding a sequence

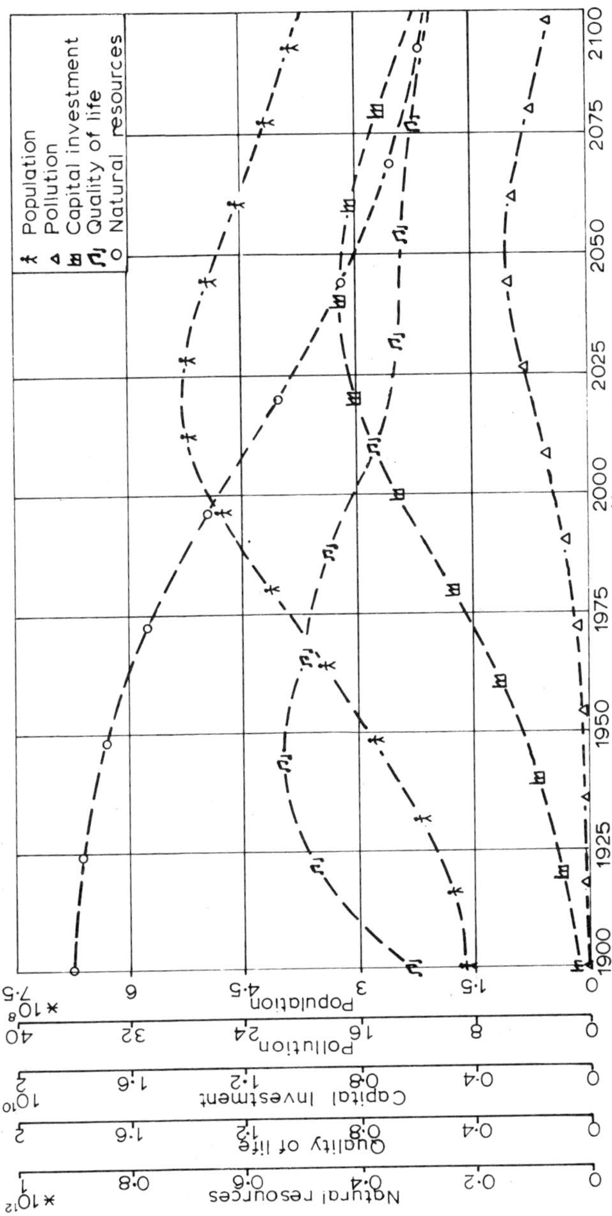

FIGURE 5.5 Forrester's World 2: a typical run from computer calculations showing the levels of Natural Resources, quality of life, capital investment, pollution, and population, from 1900 to 2100.

of overall states of the system which can be printed out from the computer. A typical run is shown in Figure 5.5 from which it can be seen that population reaches a peak and thereafter declines, natural resources remaining is eroded continuously, capital investment and pollution pass through maxima.

World 2 Relationships

The accounting transactions need no comment because they are definitionally true: if x_1 is added and x_2 subtracted from X then the new value is $X + x_1 - x_2$. The means for altering the rates merit closer attention. In the case of birth rate, we note than Malthus' necessary 'passion between the sexes' has been substantially modified so that birth rate is affected by four factors apart from the actual population size:

1 Food intake *per capita*;
2 Crowding;
3 Pollution level;
4 Material standard of living.

The sign of each causal relation assumed to operate here is that which can be generally accepted from current knowledge. It can be expected that birth rate falls under conditions of very low food intake, intense crowding or gross pollution. The fourth factor can be deduced (not necessarily as a universally applicable rule) from cross sectional data of crude birth rate per 1000 versus GNP *per capita*. The plot for virtually all countries in recent times shows lower birth rates associated with higher GNP *per capita*. There is no corresponding data to check the quantitative relations assumed for the first three factors. Forrester has made plausible guesses for the functional dependencies. This brings up an important point about World 2: many of the relationships have functional forms which are guessed rather than deduced from experiment or observation. However, this is not necessarily a reason to disregard the model as ill founded. Most models in the physical sciences contain a mixture of major and minor factors in the sense that influence on the observable effects derives strongly from the major components and hardly at all from the minor ones, so that gross errors in the parameters of the functions relating the minor components to the model behaviour can be tolerated. As an example of this effect, the relativistic variation of mass with velocity is given by

$$m = \frac{m_o}{\sqrt{(1-v^2/c^2)}}$$

where m = observed mass; m_o = rest mass; v = velocity of mass; c = velocity of light.
If v is of the order of 10 km/s (typical planetary velocity) then m is modified by a factor of about 10^{-9}. If c, the velocity of light had been mis-estimated by a factor of 10, the modification would have been 10^{-7} or 10^{-11}, but for practical purposes these corrections are all small enough to affect calculations on planetary movement very little. It is an entirely legitimate procedure to assemble a model with some numbers known only approximately in order to test the resulting structure for sensitivity of the results to variation of the guessed numbers. Indeed, there is a practical advantage in this procedure because more attention can be given to achieving higher accuracy for those parameters shown to operate with high sensitivity and less attention, or none at all, to those with low sensitivity. Given limited resources for data gathering and checking, an optimal allocation of resources can be approached through the guidance of sensitivity ranking.

For the remainder of the relations and parameters in World 2 they span a range from well known to very poorly known. Death rate is specified to be increased as food intake falls, and this relation can be defined with reasonable precision because we have the empirical evidence of starvation-induced deaths following famine. On the other hand, the notional stock of natural resources is very difficult to specify because at the ultimate extreme, the volume of say the top kilometre of the earth's crust is a conceivable limit, and yet the increasing difficulty of extracting from leaner and leaner ores puts a practical limit which is orders of magnitude smaller. Moreover, the notional natural resource stock and the manner in which the difficulty of extraction increases, turn out to be important factors affecting the results which appear in the runs.

Omissions in World 2 Structure

From what has been said about the uncertainties in parameters present in World 2, it is apparent that the run results should be treated with caution. Forrester tested some parameter variations and other experimentalists have tried many more. The results are in one sense insensitive to practically all the parameters, because ultimately the natural resource stock, irrespective of its initial size, is exhausted and collapse occurs before the exhaustion point is reached. However, the actual timing of the collapse can be varied widely by adjusting parameters. This idea of inevitable collapse brings into question the completeness of the structure because there seems to be a contradiction here. Life has existed on the planet for a period millions of times longer than the era of industrial man, and one of the reasons for this to have been possible is that the elements used by living organisms for their material existence are constantly recycled and do not become 'exhausted'. It is well known that elements of low crustal abundance, like gold and lead are already extensively recycled and it seems entirely plausible that greater difficulty in acquiring such elements will stimulate more recycling and the substitution of more abundant elements. One would have wished to see a linkage in the structure representing this recycling or substitution effect and the effect itself becoming stronger as the stress from shortage becomes more marked.

Similarly, there is bound to be some dubiety about the pollution representation in World 2 because the inhabitants do not take any steps to counter the unpleasant effects of it. Death rate rises, agriculture is adversely affected and yet no capital investment is diverted to pollution abatement. This is contrary to experience, and although there are vociferous lobbies complaining that too little is done about pollution, it is not true that nothing is done about it. The logical course would have been to insert a linkage which represented the diversion of capital investment to pollution abatement and to have made the fraction diverted depend on the nuisance level prevailing.

In the population sub-system, the mechanisms for birth rate change appear deficient in that the only feedback, apart from the purely physical ones of pollution, food intake and crowding, is from the standard of living. However, the dynamics of this feedback have been noted by several investigators to suggest the opposite effect: birth rate declines in periods of recession, when the GNP is not rising so fast or actually dropping.

Calibration of World 2

In the discussion of previous models, it has been noted that the macro properties of structures relate to the properties of the elements comprising them, so that quantitative deductions can be made in either direction. The oscillations of a spring-suspended mass can be timed and the elastic constant of the spring deduced for a known mass; or conversely, the time of oscillation calculated from the spring constant. In less well-defined structures, a test of both structural form and parameter values is provided by the demonstrated power of the model to reproduce observed behaviour. This is not a complete test because it is possible to obtain the same behaviour from more than one structure and associated 'fitted' parameters. However, the test is always useful as a means of discarding structures that cannot reproduce observed behaviour satisfactorily. Where some parameter values are subject to uncertainty it is legitimate to adjust them so that a fit between model performance and real system behaviour is reached. Indeed there are some parameters of systems whose values can only be known through this indirect fitting process. For example, Planck's constant is known via measurements on spectral lines or from black body radiation characteristics, a whole structure has to be envisaged and then h (Planck's constant) fitted to match the macro properties observed.

World 2 has been tailored through parameter adjustment to match the history of 1900–1970 for population, capital investment and food output (those variables which can be checked). This is a necessary test and it would not be worthwhile to study the model carefully unless it could reproduce some decades of history. Critics have complained that agreement over longer periods (say 1800 onwards) is not easily obtained. For high confidence one would want agreement over a longer period than 70 years.

Beyond World 2

Forrester has been at pains to point out that models like World 2 are not intended to supply point predictions, but to supply insight into behavioural modes (principally overshoot–collapse, oscillatory and gradual approach to an asymptote). Before embarking on a world model he had for some years written on the theme of the dynamic behaviour of commercial and industrial systems. He coined the term 'systems dynamics' for the particular manner of study which he employed and devised a simulation language called 'Dynamo' for modelling such systems. He argues that understanding the dynamic behaviour of systems does not come easily to the human mind, that the performance of even relatively simple models and the systems which they mirror can appear counter-intuitive. It follows from this that those responsible for the management of organizations will perform their tasks better if they come to understand, through simulations on simple models of the systems being managed, the initially surprising phenomena that are manifested. The emphasis in systems dynamics models is on the modes of behaviour that the system typically exhibits, rather than on point predictions. Modes frequently persist in spite of large changes in parameter values and hence there is no great stress laid on the accurate measurement of parameters, but much testing of sensitivity to gross changes of values. Particular sorts of mode, notably those of limit cycles or damped oscillations are found to be much more affected by the structure of the system rather than fine tuning. The presence of a delay in transmission of a feedback, for example, is a common source of oscillations. Recipes for the improvement of system behaviour are often in the form of proposed introduction of new elements or new control regimes, so altering the structure rather than changes of values in the existing structure.

At the end of *World Dynamics* is proposed a package of changes in world trajectory which would avoid doom whether from starvation or pollution. Population growth and capital stock growth are to be both brought to zero values in a short time. The continued erosion of natural resources, which would nevertheless continue, is noted and recycling receives a mention. No structural changes in which these 'reforms' would come about are offered. This is not to be construed as a criticism, but as a pointer to the need for structural change if the World 2 typical outcomes are to be avoided (assuming that its structure as in Figure 5.4 is near enough correct). The object of writing *World Dynamics* was to draw attention to the dangers ahead and thereby to stimulate the structural changes needed to avert doom. On this interpretation, the next stage would be to envisage a *Type 2* model, with a new structure emerging to supply a preferred adaptation for the human population in this finite world: Forrester's role is a catalytic one in hastening the emergence of the new structure.

World 3 and *Limits to Growth*

Professor Dennis Meadows[13] with a team of researchers at M.I.T. produced a successor to World 2 called World 3. Meadows was a pupil of Forrester's and the same modelling style is evident: the use of Dynamo, a similar list of state variables, global aggregation and a lack of restorative feedbacks which could restrain the collapse mode. However, the researchers made a concerted attempt to link the relationships which were used in World 3 to empirical evidence. For example, stocks of natural resources were adduced from the reserves of metals, etc. listed by the American Bureau of Mines. The structure is more elaborate with, for example, explicit modelling of labour in the service sector. The *raison d'être* for the model, results of its runs and a powerful accompanying message were published in a book *Limits to Growth.* This book has enjoyed much wider readership than *World Dynamics* and produced a substantial impact in most developed countries of the world. The primary message of exponential growth encountering the limits of a finite world is simple and direct, so that there has been a grass-roots response from a wide spectrum of readers. The scientific world has been more sceptical, partly because the technical report in which justification for the relationships was presented came (contrary to usual practice) some time after the conclusions had been proclaimed in *Limits to Growth.*

The structure of World 3 is so similar to that of World 2 that in the modelling sense, very little has been added to Forrester's conception. The importance of the work lies in bringing to language a new shorthand phrase 'Limits-to-growth', often in association with the 'Club of Rome', the body which instigated the work.

The Fundacion Bariloche

Among the chorus of voices raised in protest at the gloomy prognostications stemming from Worlds 2 and 3 have been people concerned that Forrester and Meadows could be right for the wrong reasons. One method of refutation of the *Limits to Growth* thesis would be the production of a single credible plan for world development which escapes the fate portrayed in the books of the first world modellers. All human decision making can be thought of as goal directed, so there is an argument for making the goals explicit and devising a plan to achieve them in order to discover whether the constraints present allow of the plan being executed. The term 'normative' is applied to such modelling exercises to distinguish them from those in which the primary object is to describe the operation of the system ('descriptive' models).

The Fundacion Bariloche, under the leadership of Professor Herrera,[14] have constructed a normative model; the avowed intention being the refutation of *Limits to Growth.* A little thought suggests that there are several separate goals to which mankind aspires; health, education, adequate housing and a secure food supply are some of the objectives that ought to be in the list. Because of the difficulty of incorporating multi-objective functions in their planning exercise, the Argentinian group subsumed the separate objectives under one all-embracing goal consisting of maximization of life expectancy. It can be argued that the contributions from the sub-goals of housing, food, etc. are necessary for the achievement of the grand goal, and that it is justifiable to ignore collinearity in deducing the relationships between each contribution and the effect on life expectancy.

The advantage of defining a single objective is that the planning procedure can be cast in the form of a dynamic programming exercise. A range of constraints must be defined: size of labour force, rate at which labour transfer between sectors can occur, land area, etc. Whereas the usual application of dynamic programming maximizes, or minimizes, some single overall property of the time span (such as cost of inventory holding during one year in an ordering and stocking context), the problem in this case is complicated by the fact that life expectancy is a variable dependent on the previous history of the individuals; it would not be satisfactory to take an integral of the product of people and their separate life expectancies over the period investigated. An additional constraint is, therefore, imposed to the effect that the solution trajectory must contain maximal life expectancy at every point of time along that trajectory. This results in a unique solution, irrespective of the stopping point.

The Argentinian model, at its first public exposure, was disaggregated effectively to four continental regions, and no explicit connections via trade, aid or migration were incorporated. Although the necessary improvements for approaching the goal are difficult in Asia, none of the four regions suffers a collapse of the Meadows–Forrester type. Thus, the exercise can be considered a refutation of *Limits to Growth*. However, the question of practicability of implementation hangs over the Bariloche model. It may be that new institutions and new procedures for planning would be essential for the materialization of these desk plans. This aspect has not been tackled by the team and indeed, the form of the model pre-supposes a high degree of centralized control, so that the study of present world conditions by the development of this model is difficult to envisage.

To see the Bariloche work in the context of the previous discussion of model types, it should be noted that here is neither a *Type 1* nor a *Type 2* model. Instead, the practicability from certain aspects of an alternative world system is being offered. Assuming that a changed structure is desirable or necessary, then for it to come about through advocacy and consensus, its emergence will be preceded by such exercises as that of the Argentinians. To sketch a detailed transition in a *Type 2* model, one would need to specify the influences for structure change which start as potential plans for alternative futures. The contemplation of these plans is the mental turmoil preceding the physical turmoil of changing institutions. The Bariloche study is not a recipe for new institutions, but rather a projection of futures which could materialize given control agencies working to the goals pre-defined.

Mesarovic–Pestel

Yet another approach to world modelling has been devised by the teams at Case Western and Hanover under Professors Mesarovic and Pestel. Their world is disaggregated to 10 regions which interact through trade. A whole suite of inter-connected sub-models (economic model, agricultural model, etc.) have been built to enable simulations to be carried out, and alternative strategies tried, by particular national decision-makers. The enterprise has as its aim, the provision of a tool for use by decision makers so that they may better understand the detailed consequences of specific policy choices. This is a laudable aim, but one would wish to be able to assess the soundness of the underlying individual models to judge the usefulness of the tool. Unfortunately the quality of transparency which was present in the early models is not evident in the Mesarovic–Pestel work. Models of any system intended as additions to the armoury of thinkers about those systems, should be open for replication and testing by other students in the field. It is by this means that the corpus of science retains its soundness.

The book which has followed the model: *Mankind at the Turning Point*[15] is, like *Limits to Growth*, strong on warnings of calamity and short on technical material by which the arguments for heeding impending calamity can be understood. Indeed, one of the warnings in the book is concerned with the misguidedness of reliance on nuclear power rather than solar power for the successor to fossil fuels. Now there may be a good case for this opinion but the modelling work has not been concerned with its investigation.

The idea of a man–computer interaction to obtain a fruitful exploration of possible choices has been applied successfully in other fields (such as design of engineering and architectural structures), but its use in this context raises awkward questions. Whereas the policy choices of the man with the model are regarded as free, or at any rate relatively unconstrained, the rest of the world must be programmed in its response. for otherwise there can be no search for a best strategy. Conceptually, this problem is soluble by a games theory attack, but developments in games theory do not yet allow of unique solutions for *n* person non-zero sum games.

Nevertheless, the Mesarovic–Pestel work offers the potential for discovering the limits which constrain world futures; given the boundaries imposed by traditional institutions and the prescribed roles played within those institutions. If the limits so found appear unacceptable, then the incentive is created to devise new machinery which would enable the limits to be exceeded. The conceptual framework is thus pushed beyond that of the Bariloche model, which has not been designed to explore the means of moving from the present to a more desirable future, but only to deduce theoretically optimal pathways.

MOIRA

Though not strictly a world model in the sense that the previous attempts can be described, the MOIRA[16] project of Professor Linnemann and his team at the Free University of Amsterdam, is a venture with the potential of expansion to full scale. Only food supply and demand is considered, all other goods and services being aggregated. Population growth is taken as an expected doubling and the primary objective is an examination of the difficulties of feeding the extra people. Disaggregation spatially is taken right down to country level, and this makes the computer runs necessarily expensive because of the large volume of data being handled.

This is a descriptive model with a strong vein of economic mechanisms running through it. The actions of the role players (farmers, consumers, government agencies) are assumed to be in accordance with economic descriptions of the collective behaviour of such classes of actor. A high degree of matching between real world behaviour and model performance is therefore possible by calibration procedures, because the economist's descriptions are quantitative in form.

The model has been used to test some of the imaginable changes that could occur, such as the abandonment of meat eating by the West. (About 40% of the non-rice cereal of the world is fed to livestock, and the energy loss via the meat route is about five-sixths of the cereal energy content.) The principal finding of the MOIRA modelling project is that the price of foodstuffs is likely to rise, in real terms, by a factor of about 3 during the period up to 2010. One aspect of real world texture which has been treated in some detail is the distribution of income. The model results indicate that in the absence of redistribution towards greater equality, the nutrition level of the poorer groups within countries will worsen.

There is a strong vein in the Dutch work, of stressing the counter-intuitive aspects of some 'common-sense' solutions, which is reminiscent of Forrester's observations on systems. For example, MOIRA has been used to show that a massive distribution of food to third-world countries would have one undesirable outcome, a depressing effect on local agriculture in those countries.

SARUM

A set of tools for world modelling has been devised by the Systems Analysis Research Unit in the Department of the Environment of the British Civil Service. The collection of structural elements with the title SARUM[17] is designed for disaggregation to any desired level in terms of the sectors of production and of the regions into which the world can be divided. The mechanisms used in the model are essentially economic in form, as the MOIRA project. Consumers are assumed to maximize utility, producers seek optimal combinations of the factors of production, etc. The effect of increased exploitation of the world's resources of land, water, minerals and fossil fuels is represented by 'depletion functions' which relate the cost of each further increment of exploitation to the quantity remaining.

There is one feature of the SARU work which indicates the possibility of representing the richness in variety of behaviour shown by the subsystems of the world. This is the notion of the 'seed sector'. In addition to the set of sectors selected to mirror activity in the contemporary world (cereal growing, coal mining, electricity generation, etc.), a latent sector can be envisaged whose general characteristics can be foreseen, but whose manifestation awaits a suitable economic climate in which it can flourish. With hindsight, this phenomenon of new sector emergence is easily understandable: the substitution of wood by coal, of sailing ships by steam, of mechanical calculators by electronic devices appear as necessary and almost obvious after the event. Given the assumption of economic forces driving substitutions, then the onset of a new sector can be modelled, provided that the vector of inputs to the sector can be specified. SARUM has been used to model the emergence of a sector manufacturing single cell protein (and competing with other animal feeds). In the early part of the run, the presence of the SCP sector is undetectable from the behaviour seen in the output variables, but as the price of feed rises, the new sector blossoms from its seed and plays a significant part in the subsequent trajectory.

This idea is capable of generalization to more aspects of a model than the rise or decline of sectors. There are structural features that can appear and disappear during the historical flow. Food rationing in Britain appeared during the last war and disappeared some years after its end. The entity termed the League of Nations did not exist (nor was there anything like it) before 1919. There is no reason why the emergence of new structural forms should not be examined by modelling methods. Although at first sight this may appear to be an exercise in the free flow of the imagination, there are severely constraining ground rules. A seed sector must be specified in terms of the vector of its input. Inventing seed sectors that need suspension of physical laws for their operation is mere science fiction. Similarly, new institutions requiring roles that could not or would not be performed is a futile exercise. However, within the ground rules, there is scope for a range of futures and to envisage even a part of this range is clearly a useful enterprise.

Comparisons between models

Different ways of world model building have been considered and at this point it is appropriate to attempt some categorization. Three styles of model representation have been used in this and the preceding chapters:

1 The Causal Link Diagram

Figure 2.2 shows the variables with which Meadows' hog-cycle model deals, drawn to depict the influence which each has on the others. The presence of a linking line means that some influence exists and absence of a line implies an insignificantly small influence. The arrow indicates the direction of causation and the sign shows the type of influence (increase of B accompanying increase of A is positive, while decrease of B accompanying increase of A is negative). Such diagrams convey the bare essentials and only broad statements can be made about behavioural characteristics.

2 The Flow Diagram

Figure 5.4 contains more information than a causal link diagram because the symbols and link types have been designed to distinguish functions. Levels and rates have the symbols respectively of rectangles and valves. The solid lines show passage of material and the dashed lines the flow of information. Such schematic ways of representing elements are useful for conveying model concepts at a depth of detail still short of the mathematical specification.

3 Mathematical Specification

The example was given in Chapter 1 of the mathematical specification for the pendulum, the spring and the oscillatory circuit. In Chapters 2 and 3 further specifications for models were quoted. The advantage of the mathematical notation is that it can be complete and unambiguous. The disadvantage is that the meaning only becomes clear by working through examples or by familiarity with particular forms, rather in the same manner as one acquires acquaintance with oft-used phrases in a spoken language. There is a further limitation in that some sorts of behaviour are not readily amenable to mathematical description, in terms of ordinary functions. New notation can be devised but this makes the description more esoteric and defeats the purpose of easy communication.

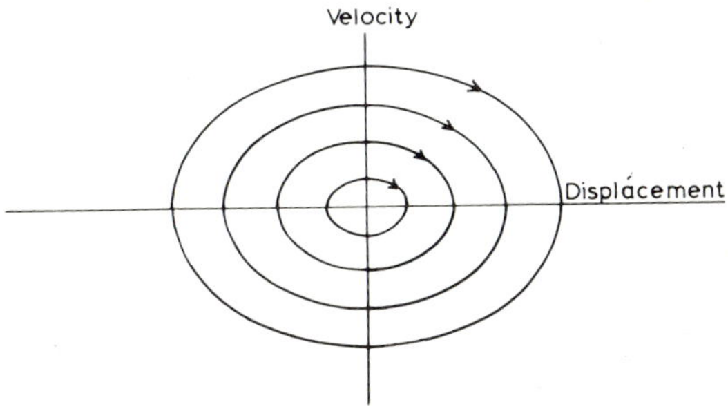

FIGURE 5.6 Representation of a model through a 'field' diagram. Here the field for a spring-suspended mass is depicted graphically.

A further means of representation is possible through 'fields'. Given a space of as many dimensions as the number of variables needed to specify the state of the system, then points in this space can be linked to show the passage of the system through time. Thus the trajectory of the system is defined for any prescribed initial state. An example of a field is shown in Figure 5.6 for a spring-suspended mass. The two variables defining the state are the displacement from equilibrium and the velocity of the mass. The field lines are ellipses but there is no particular significance attaching to the major axis being along the distance axis rather than the velocity axis, this depends only on the scales chosen. Time itself is not one of the state variables.

It can be seen that the field provides a comprehensive account of the system, unrestricted by difficulties of mathematical representation (the field lines can show discontinuities). In Figure 5.7 and 5.8

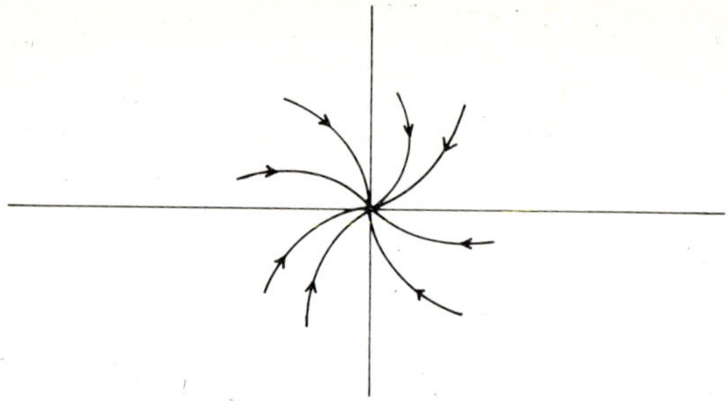

FIGURE 5.7 A field diagram of the same spring-suspended mass but immersed in treacle.

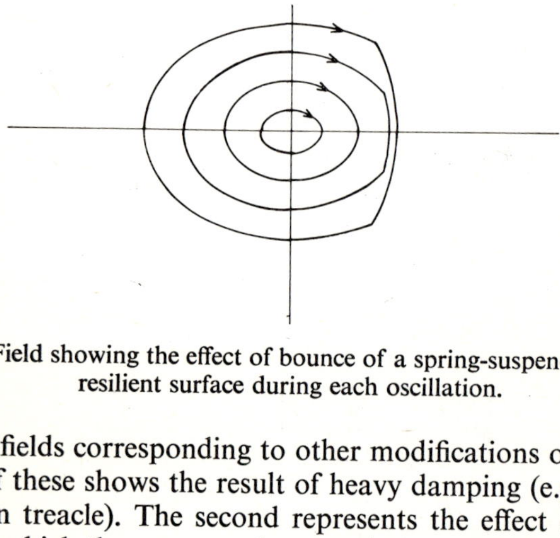

FIGURE 5.8 Field showing the effect of bounce of a spring-suspended mass off a resilient surface during each oscillation.

are shown fields corresponding to other modifications of the system. The first of these shows the result of heavy damping (e.g. immersing the mass in treacle). The second represents the effect of a resilient surface off which the mass can bounce during each oscillation.

All systems can be represented in this manner though if there are many variables, the problems of comprehensive graphical display become dire. The underlying credo is that repeatability is possible in principle. If all the state variables are made to take the same values as on some previous occasion the behaviour will be identical with that previously observed. The fact that complete repeatability is not possible in some instances does not invalidate the principle.

The field of a system can be mapped by putting it into separate states and observing the successor states to which it moves. To sample comprehensively is so laborious that in practice assumptions are made about the characteristics of the field from small samples. This procedure is defensible because of numerous past successes in science. If a system does not repeat under ostensibly the same conditions then it is assumed that an as yet undiscovered variable has changed in the intervening time, or that intrinsic 'noise' is present so that the field lines are fuzzy.

The place of mathematical specification can now be perceived within the context of the field representation. If the field lines are of simple forms then rules can be constructed which deduce the next increment of field line from the current position in state variable space. The rules applied to each point in variable space could be used to build the whole set of field lines (as opposed to the repeated sampling approach). The rules are often stated in mathematical form, hence the general use of the term 'mathematical model' to represent a system. The rules are sometimes more readily understood by stating them in verbal form, even though the material being handled is quantitative. For example, in economics a rule being used to describe the behaviour of a consumer, investor, etc. may be more easily comprehended by couching it in the form of goals which the actor wishes to achieve, rather than directly to what he does in order to obtain achievement.

The several approaches to world modelling can be related to the field representation. First, extrapolation methods for forecasting assume that the field lines for the world system contain no discontinuities and that the line which we currently occupy will show no sharp curvature as the state variables continue to change. Next, World 2 and World 3 assume that rules governing the field lines can be set down shortly, and persist unchanged throughout the progress of the world system for the next hundred or so years. The Bariloche approach is to define points in state variable space which it would be desirable to reach, and then to construct trajectories through time which are designed to meet the desired points. These trajectories are not the same as the field lines (state variable space does not have time as a dimension), but the rules necessary for designing the time trajectories could be used to define the field lines, even though the resulting field lines rules appear unlike those apparently in force. The Mesarovic–Pestel attack assumes that some parts of the rules can be changed by the process of the human role players devising different strategies in consequence of their pre-vision of the future. Linneman's group use the rules of traditional economics to discover likely world trajectories experimentally, some parts of some rules are changed to discover the effects. The SARU approach attempts the insertion of rule changes conjured up by the occupation of novel positions in state variable space.

The preceding paragraph is much concerned with rules. Some of the rules in a world model are more or less direct interpretations of physical law. The rest of the rules are concerned with human response to specific situations, points in state variable space. In order to even begin construction there must be the expectation that people *en masse* behave similarly in similar situations. This can be conceded at a primitive level: if people are hungry they will make strenuous efforts to obtain food. The whole study of economics rests on the expectation that, in general terms, aggregate human response is repeatable and predictable. There is a fuzziness about any response functions which may be devised to represent aggregate behaviour. Price elasticities of demand do not have the same enduring quality as specific heats of substances. However, this imprecision in describing economic man is not in fact a serious source of uncertainty in world models. Examination of the data suggests that man as consumer, worker, investor and manager (in the aggregate) is predictable enough for modelling purposes. Economists suffer an unwarranted amount of abuse for failing to be accurate in prediction, because the variables about which there is most interest, like unemployment, are the most sensitive to minor fluctuations in the economy. If they were asked to predict level of employment instead of unemployment the percentage errors would be ten times smaller.

The uncertainty in world models derives partly from our incomplete knowledge of the physical constraints, partly from the imprecise nature of human response functions, but principally from the fact that the field lines cannot be mapped completely. The reason for this is that the system has memory. Memory is characteristic of living organisms. If an animal is treated as one might treat an inanimate system like a spring-suspended mass then problems of repeatability arise. Unlike the spring, a late sequence of tests on the animal is found to be affected by the earlier tests. Pavlov's conditioned reflex is a well attested phenomenon.

One aspect of memory is learning, and in individual creatures (man included) this proceeds to some limited extent during one life time. However, in human society there is the potential for cumulative memory in the form of preserved records. To predict how a given society will behave at a point in future time it is necessary to know not only the 'external' state variables, but also the 'internal' variables which can be called collectively its state of knowledge. Returning to the classification of model types, a field with fixed lines in state variable space is a *Type 1* model and it appears from recent studies of automata that elementary learning processes can be represented in such a model. But the modelling of a machine or an organism or a society which discovers new patterns, ones that were previously unknown to the modeller, is a feat of a different order and termed *Type 2*.

It may appear, at first sight, that nothing further can be done about an incomplete field mapping, but an example building on Figure 5.7 shows that incomplete knowledge is not the same as no knowledge. Figure 5.8 refers to the spring-suspended mass encountering a resilient surface, but suppose that the information about the system was incomplete. Some possibilities are:

1 The surface has strong adhesive properties, and on impact the mass sticks to it, so that the field line ends there.
2 The surface is electrically charged to a high potential and the mass being charged similarly, repels it. The field line thus becomes distorted and at no point actually meets the surface.
3 The system contains an experimenter who, desiring to avoid contact between mass and surface, has arranged for the spring support to be moved away from the surface whenever there is a risk of impact.

It is clear that if no more is known about the system than the impossibility of the mass passing through the surface to the other side, then the class of all possible field lines is significantly constrained: they must all lie to one side of the surface. If more constraints of this type can be discovered then the class of all possible field lines is narrowed further, though still without generating a unique field line from a particular point in state variable space.

We can enjoy relatively high confidence in our knowledge of the various constraints of the physical world, within which human society will evolve. We cannot know in advance the means that will be adopted to take account of the constraints, as this would entail our knowing the learning that will take place in the intervening time, and that can only be known by living through the time. However, the class of possible adaptations is narrowed by consideration of the constraints. Apart from the obvious physical limitations on area of land surface, quantities of materials and energy supply, there are 'speed' restrictions. For example, it is not practicable to build solar collectors which would substitute for all current energy sources, in a space of 1 year or even 10 years. Given the will and the design competence, the final constraint which operates is the bit processing capacity of human beings.

Trends in world modelling

Some of the constituents of the several world models have shown convergence as different modelling groups have sought greater precision in defining the data. This effect is most noticeable for the agricultural sectors of models. The area of the globe which could be used for crop growing emerges as a figure differing little among the estimates made for it by the several teams. The quantity of food which can be grown on this land at varying levels of fertilizer and irrigation application is again not a figure over which there is dispute, as much empirical data is available to make the estimates reliable.

The fear of ultimate scarcity in supply of minerals which figured prominently in the first models has faded as more careful analysis of the distributions of the various ores reveals very large potential, given the capital equipment and energy required for exploitation.

The trend that appears strongly in contemporary modelling ventures is much less on the arithmetic of inventories and more on the organizational aspects of the global community. Scarcities are sources of stress: the stress does not always generate enlightened responses. One of the uses to which the Mesarovic–Pestel model has been put is the study of oil prices which are likely to prevail in the post O.P.E.C.-monopoly period. It seems likely that the demands on global modellers will emphasize more and more the short-term, the expedient and forecasting the outcome of contests. This development in modelling will suffer from the same snags as accompany efforts to forecast unemployment levels. By accenting the sensitive, error-prone variables, the focus is on failures and the modelling process is dismissed as unproductive. This would be like abandoning global climate modelling because accurate local weather forecasting is not yet possible, and similarly misguided.

6. The scope for world models

After the short review of world modelling, it is appropriate to go back and consider the fundamentals of this activity. There are some writers who have doubted the usefulness of it; it seems that these doubts are directed more at the techniques employed than at the underlying idea of taking stock of the present position. Only the most feckless deny the wisdom of taking stock. However, it appears that the sense of responsibility for avoiding short-falls and crises diminishes as the setting is larger. Individuals take trouble to balance income and expenditure, to seek insurance for their jobs, their houses, their old age, and their dependants; to plan for the future by saving and educating their children. As managers of enterprises they attempt to secure continuity of the work force and of the viability of the company. As elected representatives they try to plan for a future free from the trauma of recession and conflict. But with increasing scale and longer time-horizons, the feeling of responsibility diminishes and a Micawberish attitude supervenes. The ocean is vast and who feels responsible for it? Whereas the householder usually cares for his garden and will take action to avoid its desecration, people do not stir themselves much to avoid insult to the ocean in the form of noxious discharges or overfishing. It seems that the larger the setting the more inclined people are to relegate the reversal of decline and degradation to their descendants. Carbon dioxide concentration grows in the atmosphere as fossil fuel is burned in increasing quantities. There is no doubt that a deleterious level can be reached, but this problem is so big and diffused that people turn away from it and busy themselves with smaller matters.

It will be granted that not everyone can be sensibly concerned with all the problem areas; that there should be division of labour, and that there should be smaller efforts spent on the study of remote contingencies than on those which are pressing. Perhaps 400 researchers are currently involved in remote contingency matters. Out of four thousand million people, that is 0·000 01%: hardly excessive.

Foundation

The compilation of inventories is a sensible starting point. We should know how many more people can be fed from the earth's surface using photosynthesis. Current growth rates are also important because the categorization of the contingency as one of 10 years away, 100 years, 1000 years, etc. can follow. Possible rates of change of growth rates matter because stock for a coverage of 100 years is no comfort if the rate of usage cannot easily be shifted by more than a few per cent. The actual current rate of acceleration is less interesting: any physical exponential process becomes absurd in time, and the calculation of climax figures from hypothetical exponentials is useful only for capturing the attention of an audience. The speaker points out that a population increase of 2% p.a. means a doubling in 35 years and at 3 doublings per century, just one millenium would suffice to reach 1000 people per m^2 over the entire earth's surface. Startling though it is, this calculation is less relevant than the records of how fast birth rates can change, or rather have changed. Fall rates as high as 3% p.a. have occurred in this century. If this rate were widespread we would not reach even the first doubling.

Model form

The study of history indicates that societies devise means of adaptation when environmental constraints change. Unlike many of the earth's fauna, mankind is omnivorous so that a scarcity of one sort of food will result in shift to another. Fuels are substitutable and most artifacts can be constructed from a whole range of materials. Thus the questions to be answered by a model are first, whether the adaptations that can occur in the world system are acceptable and secondly, whether the adaptations that do occur will be acceptable. The modelling process is neutral, but what is considered acceptable depends on the culture in which the assessment is made. The adaptations by which animal populations adjust to their environmental constraints often involve a rapid thinning of numbers through starvation. Although there have been human societies in which this means was considered fit and proper, we consider it preferable to seek alternative forms of adaptation. It is clear that states of the system are possible in which none of the adaptations that could occur are acceptable. For example, a run down of food stocks followed by crop failure leaves no desirable choices. Additionally, there can be cases arise in which two or more adaptive modes are possible and there is a preferred ranking, but the adaptive mechanism in force steers towards an inferior mode. For example, there can be substitution opportunities where the declining material becomes rapidly costly in its exhaustion phase and the substitute has a substantially greater cost per unit than that of the original before its decline. It can be shown that under such conditions only a low discount rate (lower than that prevailing in the economy) applied to development of the substitute can avoid a temporary fall-off in the joint supply of the two materials. Thus the actually occurring adaptive mode is likely to be inferior because there can be no incentive for the entrepreneur to employ a low discount rate when an alternative investment would yield him a higher return. This instance has similarities with the 'commons paradox'. A resource not subject to ownership, like ocean fish, will be exploited to extinction because it is not to the advantage of any fisherman to hold back, and if a quotas agreement is made it will be to the advantage of any given fisherman to renege on it.

These two types, of what one could term 'failure of adaptation', provide a justification for attempting models which probe beyond the mere extrapolation of trends.

Learning

There is a natural desire on the part of modellers to have their models influence policy; clearly the intention of the first world modellers was to produce sufficient impact on readers to steer the climate of thought in the direction of awareness of limits and the need to change life style in an anticipatory adaptation. However, there is no necessary connection between the contemplation of a model and the policy which ought to be selected. The reason for this lies in the difficulty of distinguishing the 'greater good'. There are circumstances in which allowing people to make mistakes is a good idea. A teacher will create conditions in which pupils make mistakes in order that the impact of the errors stimulates the learning process. A strategy of evolutionary operation has been proposed for entities like chemical plants. This strategy consists of permitting or introducing noise into the values of the control variables so that by monitoring the outcomes, knowledge is gained of the potential for improvement: knowledge not accessible by deduction.

The teacher knows that if the pupil solves problems by a taught procedure—a set of rules to be memorized—then although examinations may be passed, the goal of kindling understanding has not been achieved. On the other hand, a completely *laissez-faire* style of education in which all learning occurs by chance discovery (a sort of continuous serendipity) is very ineffective. A judicious interplay of spark-kindling and vessel-filling is necessary if both informed and inquiring adults are to emerge from the process of education. There is a further material point for consideration. To allow students the freedom of the laboratory in order that discovery be uninhibited would bring the dangers of poisoning, electrocution, infestation and conflagration.

The learning process is apparent in societies, with a higher proportion of experiment and chance discovery (as opposed to guidance) than occurs in academic institutions. There are proponents for more guidance (authoritarian regimes) and there are those who argue for more experimentation. There can be no resolution of these conflicting views without agreement on the 'greater good'. The travail of learning in a classroom is justified by enlightenment that follows, but drawing up an account balance for changes in society whether technical or political is far from straightforward. The enigma of the social balance sheet has a direct bearing on the relationship of models to policy. Models can be used to disclose the class of possible adaptations, but the modeller who urges a particular course favouring one of the class has consciously or unconsciously chosen the ranges for some of the crucial weighting coefficients in the social utility function.

To put some flesh on the bones of this argument, consider the adaptation to slackening of the world's oil supplies. There has been much energy modelling: tracing out scenarios, many of which contain massive development of fission power. The class of possible adaptations contains the following elements which, present to varying degrees, defines the range of outcomes:

1 A rising share of fission power;
2 A rising share of non-oil hydrocarbon fuels (principally coal);
3 A rising share of non-fossil and non-nuclear energy sources (solar, wave, wind, geothermal, tidal, bio-fuel, ocean thermal gradient);
4 A greater thrift in energy use;
5 A fall-off in standard of living.

The modelling of the adaptive processes can be carried through with very many of the numbers known well enough for the exercise to be useful. However, there are some parts of the scenarios that are vague. The scale of deaths, accidents and environmental insult following widespread use of fission power is not known. The manner in which the rich countries would settle to a low energy income is not known (the power of habituation is, in retrospect, astonishing). If high weighting is placed on prudence and the avoidance of unknown risks then eschewing of fission power follows, because the risks associated with the other courses are relatively well understood. Now the other dimension from the learning process merits consideration. The dangerousness of a technology comes to be known, like any other hazard, by the actual events manifesting: this suggests that learning of nuclear hazards will happen through experience (perhaps *should* happen through experience as the only sure learning method). Countering this is the teacher's stance, pupils should encounter only mild risks in their experimental work because the end benefit of the learning is lost if the experiment is fatal. One more point is worthy of note. In the United Kingdom the road accident fatality rate has reached about 7000 p.a. If the consequences of motor car development could have been foreseen in advance, would we have envisaged the amazing degree of habituation to this toll? After such an experiment, we should be prepared to envisage a world accepting a large and even a rising toll from nuclear spin-off with equanimity. Equally we should be prepared to envisage a world adjusting gently to modest energy flows *per capita* (say 1 kW per person, compared with U.S. 12 kW per person) without any fuss.

Morality

There are informed people on each side of the debate about the pros and cons of nuclear power; there has been sufficient exploration of the factual content and the ethical issues to doubt that a consensus will arise unless some startling new fact appears. The modeller's role in this matter is to compose his scenario constructions with as much verisimilitude as possible; to explore all the principal members of the class of conceivable adaptations, reaching for as much precision as the data will allow. If he espouses a particular course of action, or biasses his explorations so as to emphasize one course rather than another, then he courts loss of credibility. The usefulness of a modeller who has been shown to have distorted his work, is impaired in a similar way to that of a scientist who fakes his results. This is not to imply that the modeller should not have opinions of his own, but that his opinions and his professional code are two separate things (as they are for other professions like law and medicine).

Another facet of role definition deserves comment. Whereas the stance of the neutral modeller which has been described above is impeccable, the association of the modeller and his output with a given institution known for a particular attitude, inevitably colours the judgement of critics who assess the objectivity of that output. There is no single answer to the dilemma that appears. Either the modeller chooses to sever his relation with the institution or he decides that the greater good will result from his pushing from the inside to alter the institution itself.

The modelling spectrum

A model is a way of thinking about something. New models are interesting because they offer new viewpoints, new ways of thinking about things. In this sense the creation of a model has something in common with art. The artist offers us a new vision. One can enjoy the elegance, the economy and the coherence of a model in a similar way to the appreciation of a painting, a musical composition or a poem. The contemporary fashion favours models that have obvious utility. Operational researchers stress the practical character of their work. In modern China there is a similar stress on the need for art to illustrate and enhance the socialist philosophy. From the West, this pre-occupation with worker solidarity and the creation of the beehive state seems somewhat grotesque: doubtless in another time our own insistence on the practical application of science will also appear bizarre. The usual distinction made between the manifest influence of science and engineering on the texture of life and the decorative, flimsy ephemera of art on the other is suspect. Plato viewed the activities of poets and artists as dangerously subversive: in recent times we can speculate on the connection between those artists' impressions of the world to come that appeared in the '30s with the creations all about us, from the minds of engineers whose formative years were spent exposed to the '30s science fiction.

Even the class of models which could be called useful do not necessarily result in overt changes. Some superficially chaotic systems are found, on deeper analysis, to possess subtle self-regulatory properties. There was no interference before the analysis because no-one knew how to interfere effectively, and after the analysis there is seen to be no need to interfere. If we move inwards to the subset of models with an end use, there is a spectrum which corresponds to the degree of approval from the onlookers. Consider a model of the energy flows for a house. In the temperate zone most of the input flow goes to space heating and all of the output goes to warm the environment. By a study of the thermal conduction in the various parts of the shell and a little materials application to reduce conduction, it is possible to alter the relation of the energy flow level to the temperature difference between inside and outside the shell. This can result in substantial fuel savings at quite low cost and meets with near universal approval (one supposes that fuel salesmen have reservations). In a different context, the flows of parts in and out of a store within a factory making assemblies can be modelled. For some years now there have existed well defined procedures based on stock control models, which applied to this type of situation will usually yield the benefits of lower stock holding cost and fewer stock-outs of parts. Insofar as the various people responsible for the store are concerned, this result is often a mixed blessing. It illustrates their incompetence and as a result of the change their work style becomes restricted by the enforcement of the procedures. Everyone else, the production force, the accountants, the sales force, the customers and the shareholders all approve.

For the third example, consider the modelling of alternative road systems in order to quantify the benefits resulting from additional roads. The transportation models take into account the various sorts of road users, distribution of vehicle routes before and after the change, the time saved (which is the chief benefit) and of course, the cost of the scheme. An impressive body of studies has been assembled in this field over the past decade. However, the actual implementation, or expressed intention of implementation of road schemes in recent times has met with considerable vilification, not only from the people directly affected but from many of the onlookers.

At the level of managing the entire economy, it is taken for granted that whatever policy is currently being applied will meet with disapproval from sizeable sections of the community. It seems unlikely that anyone coming forward with a systematic account of the way that the total world system works accompanied by advice on what nation states should do individually and collectively will be greeted enthusiastically. However, there is a certain timeliness in the present milieu for large system modellers.

The future for world models

The learning process depends largely on error signals and partly on the imaginative structures created to cope with the perceptions of the new situation. The error signals are necessary for the stimulation of the imagination. It needs a fuel price increase to start householders thinking about thermal insulation. Stock control experts appear when inventories are judged to be high. The signals coming from the planetary state (scarcities, rising pollutant levels and economic disturbance) are stimulating the imaginations of commentators. Because adaptation is necessary, this is a period in which societies are receptive to novel ways of thinking about the situation. There are opportunities for presenting a range of options and world modellers have a part to play because a large canvas is needed to make considered judgements of the options. Irrespective of the moralities local to time and place, the same final judgement applies to all species: adapt and survive or fail to adapt and perish. Modellers can quantify the constraints within which the adaptations must occur. They can go further and discover those adaptations which do least violence to the life styles which have evolved over millenia.

In each society it is a brief period during which these imaginative modelling ventures can have influence. While the signals are weak there is no reason to listen to warnings, and after a short interlude of awakening and contemplation of the array of possibilities, the signals become so strong that only the short term matters, only immediate survival is the issue and the luxury of objective long-term appraisal is abandoned.

If this were the whole story, then the utility of attempting influence could be questioned, but there are enough examples of thinkers who steer the course at critical periods to believe that the effort to see clearly, and communicate the picture, is worth while.

World model development

The modelling fraternity have chosen to label the members of the series of world models so far developed as either normative or descriptive. The Bariloche model epitomizes the normative approach and the SARUM structure that of the descriptive style. The extremes of either type founder in paradox. From the normative view a better world can be planned, but clearly the starting point is worse than if a normative plan had been implemented from a previous time point, and one suspects that at a future point in time there will have been no improvement but another normative plan can be offered. It is always jam tomorrow, the whole exercise begins to appear as a special variety of wishful thinking. In contrast, the descriptive model is steeped in the world as it has been and as it is. Because its descriptions include the decision makers, it ends by leaving no freedom for any significant departure from those quantified descriptions of where the world will go (apart from the cumulative effect of random error). The better the description, the less freedom that actually exists for variation on the pre-determined pathways.

The normative modellers tend to be called unrealistic and impractical: descriptive modellers run the risk of being classed as fatalists. Interesting and possibly catalytic models will lie between the extremes. A normative model in which *all* of the constraints have been inserted will begin to look much like a descriptive model but with potential for change. A descriptive model applied to historical time series will disclose points where prediction (even statistical prediction) does not generate the actual branch path taken. There have been turning points in the past and there will be future turning points. After all, the consistent fatalist modeller would have to believe that all of his models were themselves predictable and that idea contains an inherent contradiction with no resolution.

The next generation of world models

The elements of the current generation of world models can be listed:

Production functions;
Demand functions;
Income distributions;
Input–output matrices;
Investment mechanisms;
Fertility and mortality mechanisms;
Prices, wages, markets and trade;
Constraints on supply of labour, raw materials, land, water, energy, absorption capacity of the environment;
Technological change;
Structural linkages that knit the pieces into a dynamic whole;
Sets of groupings—countries, continents, races, religions, rural–urban dwellers, cartels, etc.;
Hierarchies to represent levels of power;
Policy variables;
Social indicators—quality of education, housing, health, etc.

Not all of the models have each of these elements present. Worlds 2 and 3 use only a single population with no differentiation into regions or blocs. The Bariloche model has no trade flows among the four continents.

There is a natural tendency for modellers to attempt extending the range of elements in their models so that more of real world texture is caught. However, this process is accompanied by greater difficulty in establishing validity for the additional relationships needed, and by greater problems of comprehension for both the model constructors and the audience to whom the model is presented. It seems very likely that a parallel development will occur of combining detail in one area of the model with a more highly aggregated context surrounding the spotlight. The human brain has evolved a similar tactic for taking in the information from visual signals. Only a small spot on the retina has the high density of receptors necessary for the perception of fine detail: the surrounding area delivers a much vaguer impression to the brain. We examine elaborate scenes by a serial scanning of the detail; moving the spotlight and giving more attention to the salient features.

Among possible new elements that could be considered for introduction are:

Conflict;
Degree of centralization;
Religious movements.

Conflict of various sorts between ethnic groups, religious groups, nation states, etc. has been a characteristic of human relations for a long time and is arguably as important as scarcities, technical changes, etc. There have been conflict models constructed but none of those published have provided a world model setting in which the drama is enacted.

Degree of centralization of political authority and of production has received considerable attention by commentators in recent times. Among the possible variations of life styles that could come to pass, a strong shift to decentralization and more localized living has many ramifications that could be explored effectively with world model structures.

The underlying beliefs and aspirations of people have clearly had a profound effect on the sorts of societies which have developed. Current world models have taken the mainspring for granted, but in view of the power of cults and ideologies, the only excuse for neglect is the difficulty of thinking about this sphere in modelling terms. If we were able to perceive when and how the next widespread enthusiasm will arise, this might dwarf the present-day concerns.

Finally, there is always the potential for new techniques and new concepts in the modelling field itself. Catastrophe theory and fuzzy sets have their proponents. The limits to our computation power have receded so fast in the last decades that one hesitates to suggest where the bounds will lie for future models.

Appendix 1

Distribution of molecular velocities

A closed vessel insulated from the rest of the world (no flow of energy through the walls) contains a gas consisting of a very large number of molecules.

If the vessel is thought of as divided into a large number of separate cells of volumes $\omega_1, \omega_2 \ldots \omega_z$, the molecules could have been introduced by flinging them successively into the vessel so that each one occupies a given cell; ultimately the cells contain numbers of molecules $n_1, n_2 \ldots n_z$. What is the most probable distribution of the molecules among the cells? We can express the conditions:

$$n_1 + n_2 \ldots + n_z = n \text{ the total number of molecules}$$

If the ratio of the size of a cell ω_k to the whole volume ω is given by $g_k = \omega_k/\omega$ then

$$g_1 + g_2 + \ldots g_z = 1$$

To consider the number of ways of obtaining some definite distribution, it is evident that permuting the individual molecules among themselves yields the same distribution, and the number of these permutations is $n!$ However, permuting all the molecules which occupy a given cell does not yield a new realization so the number of possibilities of realization is given by

$$\frac{n!}{n_1!\, n_2! \ldots n_z!}$$

For the probability of the distribution, this number of realizations must be multiplied by the *a priori* probability of there being n_1

molecules in the first cell, n_2 in the second and so on. The distribution probability is thus:

$$W = \frac{n!}{n_1!\, n_2! \ldots n_z!} g_1^{\,n} g_2^{\,n_2} \ldots g_z^{\,n_z} \tag{1}$$

Using Stirling's theorem, which for large values of n gives

$$\log n! = n(\log n - 1)$$

Equation (1) becomes in logarithmic form

$$\log W = \text{const.} + n_1 \log \frac{g_1}{n_1} + n_2 \log \frac{g_2}{n_2} + \ldots$$

To find the most probable distribution, the maximum value of log W is needed for variations of n_1, n_2, etc. subject to the condition that

$$n_1 + n_2 + \ldots + n_z = n.$$

Since n_1, n_2 ... are very large they can be treated as continuous variables and the method of Lagrange multipliers yields equations defining the maximum.

$$\frac{\partial \log W}{\partial n_1} = \log \frac{g_1}{n_1} - 1 = \lambda, \quad \frac{\partial \log W}{\partial n_2} = \log \frac{g_2}{n_2} - 1 = \lambda \ldots \tag{2}$$

Where λ is a constant whose value is determined by the equation $n_1 + n_2 + \ldots = n$. From Equation (2):

$$\frac{g_1}{n_1} = \frac{g_2}{n_2} = e^{\lambda+1} = \text{constant}$$

$$n_1 = ng_1 = n\frac{\omega_1}{\omega}, \quad n_2 = ng_2 = n\frac{\omega_2}{\omega} \ldots$$

Hence the numbers n_1, n_2 in the individual cells are proportional to the size of the cells, or put in other words, there is a uniform distribution of molecules throughout the vessel (the cell sizes are irrelevant).

It can be shown that the maximum is very sharp, that the uniform distribution is overwhelmingly more probable than others, and indeed experimentally, the densities of different gas samples drawn from a vessel are found to be identical within the range of experimental error.

The homogeneity consequence of postulating the gas model is not unexpected, but the same technique applied to ascertaining the velocity distribution generates an interesting result. The velocities of the molecules can be imagined to occupy 'velocity space' by drawing lines from a fixed point as origin, representing as vectors the individual molecules in magnitude and direction. The velocity space can be partitioned into cells and the question posed of the number of vectors whose ends fall in a definite cell. There are two subsidiary conditions:

$$n_1 + n_2 + \ldots n_z = n \text{ as before}$$

and

$$n_1 e_1 + n_2 e_2 + \ldots n_z e_z = E$$

which specifies that the total energy E of the gas equals the sum of the energies of the individual molecules, where e_l denotes the energy of a molecule whose velocity vector points into the cell l. Taking account of these two subsidiary conditions the probability maximum is found in an exactly analogous manner to that used for the spatial molecular distribution.

$$\frac{\partial \log W}{\partial n_1} = \log \frac{g_l}{n_l} - 1 = \lambda + \beta e_1 \; (l = 1, 2 \ldots z)$$

where λ and β are the two Lagrange multipliers
Thus

$$n_l = g_l e^{1-\lambda-\beta\lambda}{}_l = g_l A e^{-\beta e}{}_l$$

where A and β are two constants. This result (Boltzmann's distribution) indicates that the molecular velocities are distributed according to a negative exponential function. For monatomic gases, in which the energy is present only as translational movement (rather than rotation or vibration), the distribution of velocities can be shown experimentally to correspond to the Boltzmann energy distribution.

Appendix 2

Distribution of trip-making

Given a division of the urban region into zones, the number of trips made from zone i to zone j is defined as T_{ij}. In order to find the number of distinct arrangements of individuals which give rise to a particular distribution of the contents of the trip matrix (see Figure 2.1), we note that this number, $W(T_{ij})$ is the number of ways in which T_{11} can be selected from T, the total number of trips, T_{12} from $T-T_{11}$ and so on.

$$W(T_{ij}) = \frac{T!}{T_{11}!(T-T_{11})!} \frac{(T-T_{11})!}{T_{12}!(T-T_{11}-T_{12})!} \cdots = \frac{T!}{\pi T_{ij}!}$$

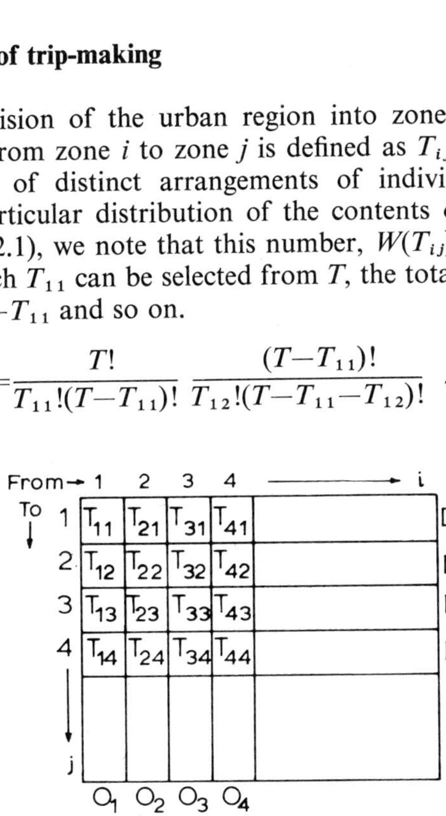

FIGURE 2.1 A trip matrix. The number of trips made in a zoned urban region, from zone i to zone j is defined as T_{ij}.

The total number of possible states is

$$W = \Sigma W(T_{ij})$$

and the constraints to which the system is subject are

$$\sum_j T_{ij} = O_i \text{ the total number of origins} \quad (1)$$

$$\sum_i T_{ij} = D_j \text{ the total number of destinations} \quad (2)$$

$$\sum_i \sum_j T_{ij}c_{ij} = C \text{ the total amount spent on trips} \quad (3)$$

To obtain the set of T_{ij}'s which maximizes $W(T_{ij})$ subject to the constraints listed, the function M has to be maximized where

$$M = \log W + \Sigma\lambda_i(O_i - \sum_j T_{ij}) + \Sigma\lambda_j(D_j - \sum_i T_{ij}) + \beta(C - \sum_i \sum_j T_{ij}C_{ij})$$

and where λ_i λ_j and β are Lagrange multipliers.

The T_{ij}'s which maximize M, and which therefore constitute the most probable distribution of trips are the solutions of

$$\frac{\partial M}{\partial T_{ij}} = 0$$

and the constraint equations.
From Stirling's approximation $\log N! = N \log N - N$

$$\frac{\partial \log N!}{\partial N} = \log N$$

and

$$\frac{\partial M}{\partial T_{ij}} = -\log T_{ij} - \lambda_i - \lambda_j - \beta c_{ij}$$

for

$$\frac{\partial M}{\partial T_{ij}} = 0$$

$$T_{ij} = e^{-\lambda_i - \lambda_j - \beta c_{ij}}$$

substituting from Equations (1), (2):

$$e^{-\lambda_i} = \frac{O_i}{\sum_j e^{-\lambda_j - \beta c_{ij}}}$$

$$e^{-\lambda_j} = \frac{D_j}{\sum_i e^{-\lambda_i - \beta c_{ij}}}$$

$$T_{ij} = \frac{O_i \qquad D_j}{\sum_j e^{-\lambda_j - \beta c_{ij}} \sum_i e^{-\lambda_i - \beta c_{ij}}} e^{-\beta c_{ij}}$$

This result indicates that a modified gravity concept in which the product of origin and destination terms yields the number of trips between origin and destination, provided that the deterrence function is an exponential function of generalized cost.

References

1. Koestler, A. (1959) *The Sleepwalkers*, Hutchinson, London.
2. Wilson, A. G. (1968) 'The use of entropy maximising models in the theory of trip distribution, mode split and route split', Centre for Environmental Studies, London.
3. May, R. M. (Ed.) (1976) *Theoretical Ecology*, Blackwell Scientific Publications, Oxford.
4. Meadows, D. L. (1970), *Dynamics of Commodity Production Cycles*, Wright–Allen Press Inc., Cambridge, Mass.
5. Mueller, M. G. (Ed.) (1966) *Readings in Macroeconomics*, Unwin.
6. Ashby, W. R. (1956), *Introduction to Cybernetics*, John Wiley, New York.
7. Ashby, W. R. (1952) *Design for a Brain*, Chapman & Hall, London, Science paperback 1965.
8. Cantley, M. F. (1973) 'Corporate planning: a review of questions and answers', Omega, *Int. J. Mgmt Sci.*, **1,** No. 1.
9. Fisher, J. C. and Pry, R. H. (1970) 'A simple substitution model of technological change', Report No. 70-C-215, General Electric, Schenectady, New York.
10. Marchetti, C. (1976) 'On hydrogen and energy systems', RR-76-7, International Institute for Applied Systems Analysis, Schloss Laxenburg, Austria.
11. Meyer, F. and Vallee, J. (1975) 'The dynamics of long-term growth', *Technological Forecasting and Social Change*, **7,** No. 3.
12. Forrester, J. W. (1971) *World Dynamics*, Wright–Allen, London.
13. Meadows, D. L. (1972) *The Limits to Growth*, MIT Press, Cambridge, Mass.
14. Bruckmann, G. (Ed.) (1974) 'Latin American world model', Proceedings of the 2nd IIASA symposium on Global Modelling, CP-76-8, International Institute for Applied Systems Analysis, Schloss Laxenburg, Austria.

15. Mesarovic, M. and Pestel, E. (1974) 'Multilevel computer model of world development system', Proceedings of IIASA symposium, International Institute for Applied Systems Analysis, Schloss Laxenburg, Austria.
16. Linnemann, H. (1976) *MOIRA—A Model of International Relations in Agriculture*, North-Holland Publishing Company, Amsterdam.
17. Roberts P. *et al.* (1977) 'SARUM 76', Research Report, Department of the Environment, London.

Index